Neuro-Rehabilitation
with Brain Interface

T0321302

RIVER PUBLISHERS SERIES IN COMMUNICATIONS
Volume 40

Consulting Series Editors

MARINA RUGGIERI
University of Roma "Tor Vergata"
Italy

HOMAYOUN NIKOOKAR
Delft University of Technology
The Netherlands

ABBAS JAMALIPOUR
The University of Sydney
Australia

This series focuses on communications science and technology. This includes the theory and use of systems involving all terminals, computers, and information processors; wired and wireless networks; and network layouts, procontentsols, architectures, and implementations.

Furthermore, developments toward newmarket demands in systems, products, and technologies such as personal communications services, multimedia systems, enterprise networks, and optical communications systems.

- Wireless Communications
- Networks
- Security
- Antennas & Propagation
- Microwaves
- Software Defined Radio

For a list of other books in this series, visit www.riverpublishers.com
http://riverpublishers.com/river publisher/series.php?msg=Communications

Neuro-Rehabilitation with Brain Interface

Editors

Leo P. Ligthart

Ramjee Prasad

Silvano Pupolin

LONDON AND NEW YORK

Published 2015 by River Publishers
River Publishers
Alsbjergvej 10, 9260 Gistrup, Denmark
www.riverpublishers.com

Distributed exclusively by Routledge
4 Park Square, Milton Park, Abingdon, Oxon OX14 4RN
605 Third Avenue, New York, NY 10017, USA

First issued in paperback 2023

Neuro-Rehabilitation with Brain Interface / by Leo P. Ligthart, Ramjee Prasad, Silvano Pupolin.

Routledge is an imprint of the Taylor & Francis Group, an informa business

Publisher's Note
The publisher has gone to great lengths to ensure the quality of this reprint but points out that some imperfections in the original copies may be apparent.

While every effort is made to provide dependable information, the publisher, authors, and editors cannot be held responsible for any errors or omissions.

ISBN 13: 978-87-7022-976-0 (pbk)
ISBN 13: 978-87-93237-43-8 (hbk)
ISBN 13: 978-1-003-33892-5 (ebk)

Contents

*Krasimir Tonchev, Stanislav Panev, Agata Manolova, Nikolay Neshov,
Ognian Boumbarov and Vladimir Poulkov*

5 An Integrated Perspective for Future Widespread Integration of Neuro-motor Rehabilitation   89

Giulia Cisotto and Silvano Pupolin

6 Ethical Issues in the Use of Information and Communication Technologies in the Health Care of Patients with Neurological Disorders   121

Matteo De Marco and Annalena Venneri

Foreword

Much attention is paid by two Conasense working groups (WG1 and WG2) active in developing a multidisciplinary vision and strategy on how Conasense can contribute to optimizing ICT in relation to quality of life (QoL & ICT). This topic is of enormous importance for society and covers many disciplines and areas of application. Selections are therefore necessary, and during 2013 and 2014, we decided to focus on those activities which should lead to breakthroughs in the sector of healthcare services made possible by integrating 3D robotics, most advanced sensing and foreseeable real-time, on-site as well as remote data processing and imaging. With some pride, we can state that until now, we are underway to develop a most interesting road map for some challenging areas of QoL & ICT. WG1 is active on the 3D robotics, sensing and services needed by the "medical world and patients", while WG2 is active on "intelligent architectures" needed for delivering real-time healthcare services and gives in this way full support to activities in WG1.

In this foreword, we summarize the major outcome of our brainstorming activities.

Members of the working groups are as follows:

WG1: CTIF Rome, CTIF Aalborg, Univ. of Firenze, Univ. of Padua, TU Eindhoven

WG2: CTIF Aalborg, TU Eindhoven, TU Delft, Nat. Taiwan Univ., Univ. of Sofia, Univ. of Hacettepe, Univ. of Firenze, CTIF Rome, Univ. of Padova, Gauss Research Foundation, Terma, CHL

It was decided that as part of a long-term (beyond 10 years) and a mid-term (5 to 10 years) road map, each WG should make a road map for a project plan for the next 3 years and give clarity about what can be reached in a period of 3 years and how finances can be organized.

Detailed discussions were held on the following subjects:

- Health care and well-being
- Flexible cooperative heterogeneous systems for enhancing quality of life
- BCI/BMI applied to rehabilitation (subject of this Conasense book)

Prime aim in brain–computer interface/brain–machine interface (BCI/BMI) applied to rehabilitation is to integrate EEG signal processing and robot device control to teach neurorehabilitation of arms and legs with expected

short-term results (1–5 years)

a) Medical service standardization
b) Integrated communication platform
c) Definition of a standardized protocol for BCI/BMI rehabilitation

medium-term results (5–10 years)

a) Worldwide standard system for medical services with international standardized phrases in national languages
b) Large use of telemedicine, teleassistance to keep the patients at home
c) BCI/BMI used for rehabilitation, prostheses, colloquia with lock-in persons
d) Digital medicine (RFID)

and long-term results (>10 years)

a) Better knowledge of brain functions and brain modelling
b) Better precise mapping of brain source signals from measurements by using EEG/NIRS and brain model
c) Use of brain model to interpret EEG/NIRS signals and help partially disabled persons to do daily jobs
d) Other simple measurement systems of brain signal different from EEG/NIRS

Inspiring fundamental discussions have been manifold and have taken place on a series of aspects such as

- Robot at home
  - System fusion
  - Real time/quasi real time
  - Data connection (pictures/video)
- Needed infrastructure
  - Choice networks and sensors
  - User interface(s)
  - Positioning services for moving patients (indoor and outdoor), robots, etc.
- Needed innovative research leading to demonstrator systems

- Services in the beginning in hospital, then in specialized centre and ending in home environments
- Support for home personal doctor
  - Home personal doctor idea
  - Robot home doctor
  - Intelligent learning robot
  - Get young scientists involved in this research

WG1 spent ample time to integrate contributions from non-tech partners (medical and patient organizations), to translate the needs in a novel approach for "BCI/BMI applied to rehabilitation" using advanced ICT to technical specifications and also to gain deep knowledge on the application. The inputs from neurophysiologists, neuropsychologists, ICT specialists on clinical data management and representatives of patient associations allow us to make this interesting and highly relevant book.

The authors have given their views on neurorehabilitation in the chapters. The material in the book is extremely valuable and useful not only for medical specialists and patient organizations but also for decision-makers and politicians. The book reflects the state of the art and gives us a view on what can be expected in the coming 10 years. We are grateful that this high-quality book can be offered to society.

Prof. Silvano Pupolin   Prof. Leo P. Ligthart
Chairman WG1   Chairman Conasense

List of Abbreviations

ADL	Activities of the daily life
ALS	Amyotrophic lateral sclerosis
AT	Assistive Technology
BCI	Brain-Computer Interface
BMI	Brain-Machine Interface
BOLD	Blood-Oxygen-Level-Dependent contrast
BRAIN	Brain Research through Advancing Innovative Neurotechnologies
CIMT	Constraint-induced movement therapy
CNS	Central Nervous System
CVD	Cerebrovascular disease
DOF	Degrees of Freedom
ECoG	Electrocorticography
ECR	Extensor carpi radialis
EEG	Electroencephalogram
EMG	Electromyography
ERD	Event-Related Desynchronization
ERS	Event-Related Synchronization
FES	Functional electrical stimulation
fMRI	functional Magnetic Resonance Imaging
fNIRS	Functional near-infrared spectroscopy
FOV	field of view
FP	Framework Programme
FRVT	face recognition vendor test
hBCI	hybrid BCI system
HBR	Human Brain Project
ICT	Information and Communication Technologies
IR	infrared
LARK	locally adaptive regression kernels
LFP	Local Field Potential
MEG	Magnetoencephalogram

MI	Motor imagery
MIF	Muscle Fibre Force
MKL	multiple kernel learning
MRI	Magnetic resonance imaging
NIH	National Institute of Health
NINDS	National Institute of Neurological Disorders and Stroke
NIRS	Near-Infrared Spectroscopy
NLS	non-linear least squares
NMS	Neuromusculoskeletal
OFL	Optimal Fibre Length
PCA	principal component analysis
PD	Parkinson disease
PnP	perspective-n-point
PTU	Pant-tilt unit
RKHS	reproducing kernel Hilbert space
S^3GP	sparse, semi-supervised Gaussian process
SCI	Spinal cord injury
SCP	Slow Cortical Potential
SDM	supervised descent method
SHREC	ShHape REtrieval Contest 2008: 3D Face Scans
SMR	Sensory Motor Rhythm
SSVEP	Steady-State Visual-Evoked Potential
TBI	Traumatic brain injury
TSL	Tendon Slack Length
VR	Virtual reality
WHO	World health organization

List of Figures

List of Tables

1

Overview on BCI/BMI Applied to Rehabilitation

Silvano Pupolin[1] and Leo P. Ligthart[2]

[1]Department of Information Engineering, University of Padua, Padova, Italy
[2]Chairman Conasense, Em. Prof. Delft University of Technology,
The Netherlands, Fellow IEEE, IET
Corresponding author: Silvano Pupolin <silvano.pupolin@unipd.it>

1.1 Introduction

Brain related pathologies both traumatic (e.g.: stroke, traumatic brain injury) and degenerative (e.g.: Alzheimer, Amyotrophic Lateral Sclerosis (ALS)), are increasing their incidence in the worldwide population affecting the large majority of old persons, thus heavily augmenting the year to live with disabilities. Today two completely different roadmaps are followed for health care of these patients: i) *medical*, which includes a wide variety of interventions from pharmacological to the new frontiers of stem cells, and ii) *neuro-rehabilitative*, aimed to expand as much as possible the functional recovery. While, the first try to repair the damaged neurons, the second try to maximise the residual functional connectivity of survived neurons to restore the activities performed before the injury. The two roadmaps are not competing each-other but aim to work complimentary to achieve the best results. In any case in order to maintain or improve patient performance continuous rehabilitation is required.

The importance of limiting the effects of damages produced by Brain related pathologies is shown by the World Health Organization (WHO) and UK stroke statistics where a wide impact of stroke on the population is shown. We are talking of about 800,000 cases in USA and 200,000 in UK every year. Seventy-five percent of the cases are for persons older than 65. Also, stroke caused 20% of deaths. Among the survivors, 64% will have no or mild

Neuro-Rehabilitation with Brain Interface, 1–8.

disability, 14% have moderate disability, and 22% have severe disability and require nursing care. These numbers show that each year in USA and UK, we have about 170,000 new persons requiring nursing care after stroke, which is about 0.4% of the whole population of the two countries.

From the above data, it appears that any time we be able to reduce severe disability due to stroke, we reduce the healthcare costs.

Hereafter we consider neuro-rehabilitation. Today neuro-rehabilitation is based mainly on direct treatment on the patient done by rehabilitation therapists (e.g. physiotherapist, occupational therapist, speech therapist) requiring patient hospitalization or at least day-hospital access for rehabilitation care. This health-care model has high costs and cannot be completely supported by the national health care systems. It is mandatory find new way for the long-life neuro-rehabilitation of these patients. This require completely new rehabilitation protocols that could be obtained by the joint involvement of specialists of several different disciplines, as e.g.: neurology, to understand the way how after a stroke neurons in the brain reorganize, how by using physical therapy we could reinforce this neuron reorganization, and how ICT could help in obtaining better recovery of the lost functions. Psychology to understand how to keep the patient aware of his/her status and how (s)he be able to accept and continue a long term and stressing therapy, how neuro-rehabilitation is, and also to accept the new ICT based therapies.

The new ICT based neuro-rehabilitation techniques require a system to measure precisely the neuron activities both in space and time and this is a real challenge to solve. Today there are experiments done in research centres which show that ICT could improve the benefit of the neuro-rehabilitation. Also, it foresee the possibility of a home neuro-rehabilitation system which could be used by each single patient at his home. In order to design this new system several engineering problems must be solved: in particular, the measuring system of the neuron activity must be non-invasive, precise, easy to use and low cost. Moreover, the system itself must have simple commands in order to be used by non-skilled operators and should have several automatic controls to avoid errors while using it which could be dangerous for the safety of patient. However, we remark that each therapy must be tested in hospitals to check its efficacy before to be released for the home application. In order to design the most appropriate neuron activity measuring system, a deep knowledge on how the brain works is mandatory. The new ICT systems for neuro-rehabilitation which measure the neuron activities and take appropriate actions are called Brain Computer Interface (BCI) in case they interact only with a computer or Brain Machine Interface (BMI) in case the interaction is done with a robot helping in doing the limb movements. In order to get the most appropriated protocol for rehabilitation we need a deep knowledge of how the brain works in order to capture the signals of interest for the neuro-rehabilitation. Some studies in this direction have been done in the past but we expect much better results by using the research results from two worldwide initiatives launched in the last years in USA and EU.

The first initiative is the **B***rain* **R***esearch through* *Advancing* **I***nnovative Neurotechnologies* (BRAIN) supported by USA [1] with the aim by 2025 to map the circuit of the brain, measure the fluctuating patterns of electrical and chemical activity and understand how their interplay creates our unique cognitive and behavioural capabilities. The second initiative is the *Human Brain Project* (HBR) [2] which is part of the FET Flagship Programme of the EU with the aim of understanding how the human brain works by 2030. The HBR will initialize fundamental contributions to neuroscience, to medicine and to future computing technology. In neuroscience, the project will use neuroinformatics and brain simulation to collect and integrate experimental data, identifying and filling gaps in our knowledge, and prioritizing future experiments. In medicine, the HBR will use medical informatics to identify biological signatures of brain disease, allowing diagnosis at an early stage, before the disease has done irreversible damage, and enabling personalized treatment, adapted to the needs of individual patients. Better diagnosis, combined with disease and drug simulation, will accelerate the discovery of new treatments, drastically lowering the cost of drug discovery. In computing, new techniques of interactive supercomputing, driven by the needs of brain simulation, will impact a vast range of industries. Devices and systems, modelled after the brain, will overcome fundamental limits on the energy efficiency, reliability and programmability of current technologies, clearing the road for systems with brainlike intelligence. The budget of each of the two research programmes is in the order of 80MEURO/year.

These two research programmes will foresee knowledge results in at least a decade from now, and the applications will follow a decade later.

Actual knowledge on how the brain works is far away from the one we expect to obtain as results from the two big programmes launched by USA and EU, but we cannot wait another two decades before to have efficient rehabilitation protocols based on the new paradigm. Surely, the research results which will come out from the two big research programmes will improve knowledge and will be used to refine what will be done.

The today's available knowledge is sufficient to understand some mechanism which operates during the rehabilitation phase in order to recover part of the motor ability, by understanding the effects of the brain damage.

It is well known that during a movement or imaging to perform a movement, some brain electric signals generated by the brain cortex reduce their amplitudes, largely during movement more than during imaging to perform a movement. However, these signals have large amplitude fluctuations during the day, so absolute values have relatively low significance. Also, when performing a movement, muscles generate an electric signal that could be detected by electromyography (EMG). Moreover, if we apply a current to a muscle, it contracts, and applying appropriate currents to a group of adjacent muscles, we could perform most of the movements we are capable to do.

Another known effect is if to a paralysed person we show a movement to perform and (s)he think to do the movement and we with a robot will perform the movement and this action is repeated many times some new neural connections on the brain appear and they begin to take care of the movement. In some cases, this is the start for the person to use his/her former paralysed limb. We distinguish the following phases of the developments in neurorehabilitation.

Firstly, the new paradigm for motor rehabilitation is to use multiple actions based on the use of the motor stimuli generated by the persons in different parts of his (her) body, to be specific the brain cortex and the muscles, related to external stimuli to perform some action. Secondly, it is used to help the movement by using a robot to apply the force that the muscle is not yet capable to do. Thirdly, it is used to give a feedback related to the movement that will be a vibration in case the muscle is activated, but the muscle force is insufficient to move the arm so that assistance from the robot is needed, or to give other visual feedback, meaning that when the movement is performed, the patient is aware of what (s)he is doing. Fourthly, it is used to give feedback when a part of the brain cortex is activated different from the part we expect to be in charge of the movement. Due to this wrong brain cortex activation, the patient may initiate movement using the robot based on a negative and unwanted feedback as, e.g., a force that is opposite to the movement.

All the above aspects are experimented today in the motor rehabilitation procedures but a full proof that they are the best does not exist. The brain studies could give us new knowledge and new options to be compared for the rehabilitation phases.

Engineering aspects to build a neurorehabilitation machine are challenging. Indeed, we need to capture brain and muscle signals, process them, and extract key features in real time to activate the feedback mechanism. In these processes, a 3D vision system is capturing the recent research interest because with a 3D system, we could infer the limb movements and find the positions of joints in the limb movements to be used in the limb model. In particular, upper limbs, especially hands, are the most challenging research activities for the complexity and the variety of precise movements they are able to do. Precise models of upper limbs and hands have been proposed to foresee the movements and they will be used in predicting the motion of the upper limb in dependence of the forces applied by the muscles. In this way, we could predict the exact dynamics of the limb on the basis of the muscles and robot-applied forces. This model will be used to design the force that the robot has to apply in order to perform the movement taking care of the patient muscles activity.

Last but not least, because we are treating persons, ethical issues are of major concern in developing new procedures for neurorehabilitation. This issue will be used for the design of a secure ICT system, regarding all the ethical aspects of the data that are stored. In any case, another accurate procedure must be designed for the persons who have access to the data for organizing the most appropriate therapies for the patient.

As it has been outlined, the solution of this problem is extremely challenging because it requires that several experts with completely different backgrounds discuss all together what to do and find the most appropriate solution for each patient. This will be a long process because it requires that each person involved be able to accept an open discussion with persons that are talking a completely different technical language. Then, the success of this initiative is related to a period in which the national health authorities will organize a continuous education programme on this topic for the doctors, physiotherapists and engineers involved.

The six chapters of the book cover several of the above-outlined problems and are an up-to-date analysis of the status of research in neurorehabilitation.

Chapter 1 by Silvano Pupolin and Leo P. Ligthart is an introductory chapter that gives an overview on BCI applied to rehabilitation.

Chapter 2 by Junichi Ushiba and Shoko Kasuga gives an accurate description of the ongoing research and experimental results obtained for the upper limb rehabilitation after stroke. It shows that brain–machine interface (BMI) will be the future revolution for motor rehabilitation.

Chapter 3 by Roberto Bortoletto, Luca Tonin, Enrico Pagello and Emanuele Menegatti shows how BMI is used to operate a robot and the way the key aspects of the BMI architecture are most effective, taking the necessity of a real-time feedback into account. Also, the development of new modelling methods and computationally efficient numerical simulation algorithms is raising the interest in musculoskeletal modelling and simulation among the biomechanical and medical communities. These models could be used for the characterization of the patient's movement biomechanics and will be used for the design of prosthesis for neurorehabilitation.

Chapter 4 by Krasimir Tonchev, Stanislav Panev, Agata Manolova, Nikolay Neshov, Ognian Boumbarov and Vladimir Poulkov is related to the 3D signal processing to extract some features regarding face orientation and emotion recognition which are helpful in all aspects where we use the system for speech rehabilitation and psychological support from remote places.

Chapter 5 by Giulia Cisotto and Silvano Pupolin analyses the actual rehabilitative techniques and the BMI systems available on the market. These systems are mainly based on two different feedbacks to the patient. One feedback uses a visual system showing on a screen what action the patient has to do and how (s)he is doing it, called virtual reality-based system. The second system is mainly used for the rehabilitation of the hand and uses electromyography (EMG) signals to detect muscle activity and give a feedback to the patient. The hand/finger movements are done by a robot. Future perspectives are to integrate the two above-listed feedbacks and sensors and to add new sensors, such as electroencephalogram (EEG) and/or near-infrared spectroscopy (NIRS) to measure the brain activity. The choice of the system to use will depend on the engineering solution and its easy-to-use. The integration of all the sensor signals is a key challenge in order to clearly obtain the features needed, both qualitative and quantitative, to command and control the robot for rehabilitation.

Chapter 6 by Matteo De Marco and Annalena Venneri is devoted to the ethical aspects which are playing a key role to keep the patient's data confidential. This appears a particularly stressed problem for persons with neuronal diseases. Relevant problems have been highlighted, and a solution from the technical and organization point of view is open.

References

[1] http://braininitiative.nih.gov/
[2] https://www.humanbrainproject.eu/

Biographies

S. Pupolin, graduated in Electronic Engineering from the University of Padova, Italy, in 1970. Since then he joined the Department of Information Engineering, University of Padua, where currently is Professor of Electrical Communications. He was Chairman of the Faculty of Electronic Engineering (1990–1994), Chairman of the PhD Course in Electronics and Telecommunications Engineering (1991–1997), (2003–2004) and Director of the PhD School in Information Engineering (2004–2007). Chairman of the board of PhD School Directors (2005–2007), Member of the programming and development committee (1997–2002), Member of Scientific Committee (1996–2001), Member of the budget Committee of the Faculty of Engineering (2003–2009) Chairman of the budget committee of the Department of Information Engineering (2014–2017), of the University of Padua. Member of the Board of Governor of CNIT "Italian National Interuniversity Consortium for Telecommunications" (1996–1999), (2004–2007), Director of CNIT (2008–2010). General Chair of the 9-th, 10-th and 18-th Tyrrhenian International Workshop on Digital Communications devoted to "Broadband Wireless Communications", "Multimedia Communications" and "Wireless Communications", respectively, General Chair of the 7th International Symposium on Wireless Personal Multimedia Communications (WPMC'04).

He spent the summer of 1985 at AT&T Bell Laboratories on leave from the University of Padua, doing research on Digital Radio Systems.

He was Principal investigator for national research projects entitled "Variable bit rate mobile radio communication systems for multimedia applications" (1997–1998), "OFDM Systems with Applications to WLAN Networks" (2000–2002), and "MC-CDMA: an air interface for the 4th generation of wireless systems" (2002–2003). Also, he Task leader in the FIRB

PRIMO Research Project "Reconfigurable platforms for broadband mobile communications" (2003–2006).

He is actively engaged in research on Digital Communication Systems and Brain Communication Interface for motor neuro-rehabilitation.

L. Ligthart was born in Rotterdam, the Netherlands, on September 15, 1946. He received an Engineer's degree (cum laude) and a Doctor of Technology degree from Delft University of Technology. He is Fellow of IET and IEEE and academician of the Russian Academy of Transport.

He received Honorary Doctorates at MSTUCA in Moscow, Tomsk State University and MTA Romania. Since 1988, he held a chair at Delft University of Technology. He supervised over 50 PhD's.

He founded IRCTR at Delft University. He is founding member of the EuMA, chaired the first EuMW and initiated EuRAD conference.

Currently he is emeritus professor of Delft University, guest professor at Universities in Indonesia and China, Chairman of CONASENSE, Member BoG IEEE-AESS. His areas of specialization include antennas and propagation, radar and remote sensing, satellite, mobile and radio communications. He has published over 650 papers, various book chapters and 4 books.

2

ICT for Neurorehabilitation

Junichi Ushiba and Shoko Kasuga

Department of Biosciences and Informatics, Faculty of Science
and Technology, Keio University, Japan
Corresponding author: Junichi Ushiba
<ushiba@brain.bio.keio.ac.jp>

2.1 Introduction

When the human central nervous system is damaged, such as by stroke or
head trauma, there are many cases where biological repair is insufficient and
chronic functional impairment remains. This not only impairs the autonomous
daily life of the person in question, but also is a large problem with respect to
the burden on caregivers and tax burden needed for medical welfare as well.

As approaches to restore function lost due to organic brain injury, there
is pharmacological and stem cell research to intrinsically and extrinsically
regenerate blood vessels, neuroglia and neurons. In addition, there is research
on neural rehabilitation to reacquire the expression of functions by reorganiz-
ing the interneuronal information transfer in the brain regions that escaped
organic injury. It is thought that such organic repair on the genetic and cellular
levels, as well as functional reconstruction methods on the cellular network
level, will be explored across disciplines in the near future and the combination
that yields the ideal interaction between them will become clear.

Research on the functional reorganization of the neuronal network has
seen rapid advances in basic science and clinical applications in recent years,
and there have been numerous attempts focused on recovery of upper limb
function following stroke hemiplegia. In this chapter, we will focus on this
point and explain these trends in research overall. Note here that engineering
solution introduced here is interpreted as an ICT, especially between brains
and machines, and could become a novel medical device to functionally heal
damaged brains.

Neuro-Rehabilitation with Brain Interface, 9–20.

2.2 Plasticity and Rehabilitation of the Brain

In one well-known study on nervous system reconstruction, an artificial cerebral infarction was induced in the upper limb region of the motor cortex of squirrel monkeys, and the reconstruction process of upper limb function was examined using microelectrical stimulation of the cortex [1]. In this study, it was shown that in the group forced to use their affected limb when feeding, the somatotopy changed significantly, and the areas surrounding the function lost to cerebral infarction compensated. In the group that used its healthy side when feeding and not the paralysed side, it was found that the upper limb control region, which was used with less frequency, showed masking by the surrounding areas' functions. This suggested that forced use of the paralysed hand prevented learned non-use, and that Hebb plasticity and use-dependent plasticity from appropriate functioning of the somatosensory feedback brought on by voluntary movement were effective at restoring function to the paralysed side. The clinical development of this is constraint-induced movement therapy (CI or CIMT). This method improves upper limb function through gradual and intensive training of the paralysed limb by restricting the healthy limb with a sling or other apparatus. This has been researched in the United States on a large scale in multicentre studies, and its effectiveness has been demonstrated [2].

Based on the constraints of its methodology, CIMT is applied to mildly hemiplegic patients with remaining voluntary motor function sufficient to carry out movement tasks. Even in cases of hemiplegia outside the scope of CIMT where voluntary movement cannot be sufficiently induced, if some degree of voluntary myoelectricity can be measured from the upper limb on the paralysed side, there are cases where movement assistance is possible through muscle contraction assistance systems [3] that provide electric muscle stimulation proportional to the amount of surface myoelectricity, and research into this is progressing for HANDS therapy [4] and power-assist FES [5]. Movement training of the upper limb on the paralysed side using these systems has been shown to induce neurally rehabilitative effects just as with CIMT.

In cases of paralysis so severe that voluntary movement and electromyography cannot be exhibited, it is difficult to externally determine the intent of movement, so it is not possible to encourage the use of the paralysed limb and its associated sensory feedback as with CIMT or HANDS therapy or power-assist FES (Figure 2.1(a)). However, even in such cases, if it is possible to utilize brain–machine interface (BMI) technology to approximate the state of the areas upstream of the outgoing motor pathways, that is to say the

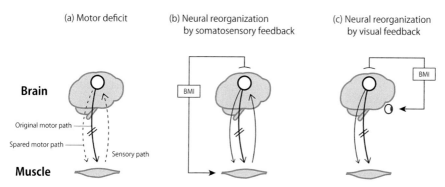

Figure 2.1 Schema of neurological conditions under stroke hemiplegia and BMI interventions

somatosensory motor cortex within the cerebral cortex, then it is possible to provide movement assistance through myoelectric stimulation or an electrical gripping device that responds to intended voluntary movement. With this, through the somatosensory motor system cycle (intention of voluntary movement, execution, and feedback of the result of the movement), relearning by the central nervous system can be expected (Figure 2.1(b)) [6].

Apart from the movement training using BMI, it is also thought that visual feedback training of the state of brain activity using BMI induces neurally rehabilitative effects. Neurofeedback training, which converts brain wave changes into bar graphs and other visual information so that they can be recognized by the subject and promotes brain wave control, is being tested as a method of explicitly teaching how to voluntarily modulate the state of brain activity for epilepsy, attention-deficit/hyperactivity disorder, autism, dyskinesia, and anxiety disorders [7]. In the field of motion control, there is known to be biofeedback therapy to explicitly relearn voluntary motion where the surface electromyograph from the paralysed muscles is presented visually [8], and a method of visually presenting motion-related brain waves in place of the electromyograph can be expected to have similar effect. The possibility that motion image training for paralysed limbs will be effective for regaining of function after stroke has long been noted [9]. However, it has so far been difficult to work with this method as it is not possible to objectively and quantitatively evaluate the state of the movement images. If it were to become possible to evaluate the state of movement images from BMI to some extent, it is believed that functional reconstruction of the motor nervous system could be actively pursued. In this manner, the concept of feedback training through visual presentation of the state of brain activity suggests the possibility of BMI inducing neurally rehabilitative effects from a different perspective than that described above (Figure 2.1(c)).

(a) **Motor deficit.** Sensorimotor cortex (shown as an open circle) innervates skeletal muscles via motor paths, but the motor paths are functionally disturbed by stroke lesion. Organically intact motor and sensory paths are then affected by the learned non-use.

(b) **Neural reorganization by somatosensory feedback.** Somatosensory sensation is artificially induced by neuromuscular electrical stimulation or motor-driven orthosis, contingently to excitation of the motor cortex. Such feedback may satisfy the principles of neuroplasticity, leading unmasking spared motor paths. Sensory stimulation itself, as well as the process of motor recovery, may also lead to functional gain of sensory paths.

(c) **Neural reorganization by visual feedback.** Explicit understanding of ongoing excitability in the motor cortex helps to learn a strategy to generate new motor commands passing through spared motor paths.

2.3 Attempts at Neural Rehabilitation with BMI

Studies to build BMIs based on the concepts introduced in the previous section with the aim of neural rehabilitation for upper limb stroke hemiplegia are being attempted in multiple locations in Japan and overseas. As methods of estimating the excitability of the somatosensory motor cortex using a scalp electroencephalogram, arch-shaped 8–12-Hz components, called sensorimotor rhythms (SMR) or mu rhythms, are often designated as the feature value. In general, the SMR is marked when at rest, and this is interpreted to indicate an "idling state" of the somatosensory motor cortex [10]. Because when interconnected activity between the thalamus and cerebral cortex increases in accordance with the performance of motion or motion imaging, the membrane potential of neuron groups begins to vary asynchronously and SMR amplitude is said to decrease [11]. This can be used to drive visual indicators, electrical apparatuses, electrical muscular stimulation, etc.

In 2008, a joint research group including the National Institute of Neurological Disorders and Stroke (NINDS) at the United States' National Institute of Health (NIH) and the University of Tubingen performed SMR measurements using magnetoencephalography and established a system for driving a pneumatic finger extension apparatus based on the results. Stroke patients who have shown SMR changes brought on by continuous use of this BMI have been reported, and the possibility of inducing plasticity of brain activity through intervention by BMI was shown [12]. In 2009, the Pfurtscheller group from Australia similarly recorded SMR in stroke hemiplegia patients using electroencephalography and developed a system to change an animated limb displayed on a computer screen in response to the change. After training over several days, SMR generated in accordance with the intent to move the paralysed side decreased significantly, cases of discrepancies in brain wave levels between rest and intended movement reaching 70%–80% were reported, and the possibility of inducing plasticity in brain activity with a BMI feedback loop was shown [13]. Following this, from 2009 to 2010, case reports of the effectiveness of BMI myoelectrical stimulation improving upper limb function [14] and case reports of improvement in upper limb function induced by combining BMI with conventional physical therapy [15, 16] have come out one after another. In addition, Singapore's A*STAR research team also constructed a system that combines BMI and robotic rehabilitation and reported the results of its implementation on 54 stroke hemiplegia patient subjects at an international conference [17]. Attempts at neural rehabilitation using BMI are gaining momentum year by year. Starting in 2011, multiple locations in Europe collaborated to start "BETTER", a stroke rehabilitation research project that combines BMI and robotics [18]. In Japan, among brain science research strategy promotion programmes started in 2008, the authors' group has been conducting research into BMI neural rehabilitation [19]. To date, for chronic hemiplegic patients, they have constructed a BMI system to aid in finger extension motion with an electrical apparatus that responds to changes in SMR generated by the intent to move (Figure 2.2) and verified its effectiveness in training (Figure 2.3).

(a) SMR. During rest, a characteristic 8–12 Hz oscillation is observed in EEG recorded over the sensorimotor cortex. The amplitude of such oscillation, called SMR, is significantly reduced during motor imagery or motor execution.

(b) BMI set-up. Motor-related power decrease of SMR is monitored from an EEG headset (Step 1). BMI algorithm judges its significance (Step 2). Neuromuscular electrical stimulation to paretic muscles, as well as motor assistance by a motor-driven orthosis, is actuated only if significant decrease in SMR is observed (Step 3a, 3b).

As a result, the fact that a decrease in SMR during the intent to move becomes significant, the fact that the at-rest excitation threshold in the primary

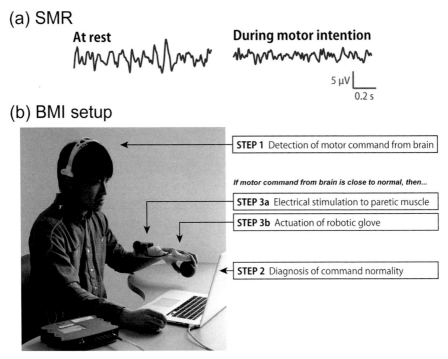

Figure 2.2　Conceptual explanation of our BMI rehabilitation system

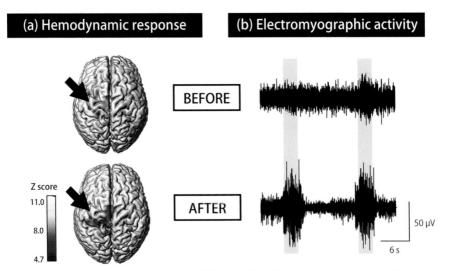

Figure 2.3　An example of BMI rehabilitation effect from a representative participant

motor cortex on the disabled side decreases and tends to improve, the fact that BOLD MRI increases in the primary motor cortex of the disabled side where there is an intent to move, the fact that observations on voluntary myoelectrograms of overall finger extension on the paralysed side improve and the fact that voluntary finger motion improves and clinical scores rise have all been shown [20–22]. In addition, from the results of one case study based on the ABAB planning method, functional improvement was not shown just by repeatedly administering stimulation to the paralysed muscle concurrently with the intent to move the paralysed upper limb, and functional improvement was shown to be brought on when stimulation was administered to the paralysed muscle only when event-related desynchronization occurred in the intent to move the paralysed upper limb. Also, from comparison of the results of different types of BMI intervention conducted at different clinical sites, it was found that stronger functional recovery occurred when somatosensory feedback, rather than visual feedback, was used [23].

(a) Hemodynamic response by BOLD MRI while attempting paretic hand movement. Red colours show statistically significant z-scores, where the task-dependent responses were seen.
(b) Electromyographic activity from the trained muscle (extensor digitorum communis muscle). Shaded time periods represent the timing when motor cues (attempting finger opening) are given to the participant. Upper trace: data obtained on the day of admission. Lower trace: data obtained on the final day of training.

2.4 Conclusions

BMI rehabilitation is at a stage where its clinical effectiveness and some mechanisms of action have been shown, and further research into these mechanisms of action is needed. Theoretical interpretation of this phenomenology, based on the neuroplasticity principle, is extremely important with respect to expansion of applicable diseases and consideration of multidisciplinary therapies involving pharmacological and stem cell treatments as mentioned in the "Introduction". In particular, the authors reference their other work examining the mechanism of action of BMI rehabilitation while integrating the concepts of neuroscience and rehabilitation medicine [23]. For example, recently, improvements in writer's cramp with BMI have been successful by deducing the framework of reinforcement learning via visual feedback, which is one possible mechanism of action of BMI rehabilitation [24]. As in this example, for various diseases that have severe disabilities with regard to motion output regulation in the central nervous system, it is believed that the applications for BMI rehabilitation will continue to expand in future. As more basic research, by measuring the internal sources of activity from the scalp with higher precision, the development of technologies that accurately decode motion-related information is proceeding, and in future, it is expected that this will be applied, leading to development in next-generation BMI rehabilitation to reconstruct articulated composite motion.

This chapter was partially revised based upon the review previously written by the authors [25, 26] and was translated in English for this book chapter.

References

[1] Nudo, R. J., Plautz, E. J., Frost, S. B., "Role of adaptive plasticity in recovery of function after damage to motor cortex," *Muscle Nerve*, Vol. 24, No. 8, 2001, pp. 1000–1019.

[2] Blanton, S., Wilsey, H., Wolf, S. L., "Constraint-induced movement therapy in stroke rehabilitation: perspectives on future clinical applications," *Neuro Rehabilitation*, Vol. 23, No. 1, 2008, pp. 15–28.

[3] Muraoka, Y., "Development of an EMG recording device from stimulation electrodes for functional electrical stimulation," *Front Med Biol Eng*, Vol. 11, No. 4, 2002, pp. 323–333.

[4] Fujiwara, T., Kasashima, Y., Honaga, K., Muraoka, Y., Tsuji, T., Osu, R., Hase, K., Masakado, Y., Liu, M., "Motor improvement and corticospinal modulation induced by hybrid assistive neuromuscular dynamic stimulation (HANDS) therapy in patients with chronic stroke," *Neurorehabil Neural Repair*, Vol. 23, No. 2, 2008, pp. 125–132.

[5] Hara, Y., Ogawa, S., Tsujimichi, K., Muraoka, Y., "A home-based rehabilitation program for the hemiplegic upper extremity by power-assisted functional electrical stimulation," *Disabil Rehabil,* Vol. 30, No. 4, 2008, pp. 296–304.

[6] Daly, J. J., Wolpaw, J. R., "Brain-computer interfaces in neurological rehabilitation," *Lancet Neurol*, Vol. 7, No. 11, 2008, pp. 1032–1043.

[7] Heinrich, H., Gevensleben, H., Strehl, U., "Annotation: neurofeedback – train your brain to train behaviour," *J Child Psychol Psychiatry*, Vol. 48, No. 1, 2007, pp. 3–16.

[8] Woodford, H., Price, C., "EMG biofeedback for the recovery of motor function after stroke," *Cochrane Database Syst Rev*, Issue 2, 2007, Article ID CD004585.

[9] Sharma, N., Pomeroy, V. M., Baron, J. C., "Motor imagery: a back-door to the motor system after stroke?," *Stroke*, Vol. 37, No. 7, 2006, pp. 1941–1952.

[10] Pfurtscheller, G., Stancák Jr, A., Neuper, C., "Event-related synchronization (ERS) in the alpha band – an electrophysiological correlate of cortical idling: a review," *Int J Psychophysiol*, Vol. 24, No. 1–2, 1996, pp. 39–46.

[11] Ritter, P., Moosmann, M., Villringer, A., "Rolandic alpha and beta EEG rhythms' strengths are inversely related to fMRI-BOLD signal in primary somatosensory and motor cortex," *Human Brain Mapp*, Vol. 30, No. 4, 2009, pp. 1168–1187.

[12] Buch, E., Weber, C., Cohen, L. G., Braun, C., Dimyan, M. A., Ard, T., Mellinger, J., Caria, A., Soekadar, S., Fourkas, A., Birbaumer, N., "Think to move: a neuromagnetic brain-computer interface (BCI) system for chronic stroke," *Stroke*, Vol. 39, Vol. 3, 2008, pp. 910–917.

[13] Pfurtscheller, G., Muller-Putz, G. R., Scherer, R., Neuper, C., "Rehabilitation with brain-computer interface systems," *Computer*, Vol. 41, No. 10, 2008, pp. 58–65.

[14] Daly, J. J., Cheng, R., Rogers, J., Litinas, K., Hrovat, K., Dohring, M., "Feasibility of a new application of noninvasive Brain Computer Interface (BCI): a case study of training for recovery of volitional motor control after stroke," *J Neurol Phys Ther*, Vol. 33, No. 4, 2009, pp. 203–211.

[15] Broetz, D., Braun, C., Weber, C., Soekadar, S. R., Caria, A., Birbaumer, N., "Combination of brain-computer interface training and goal-directed physical therapy in chronic stroke: a case report," *Neurorehabil Neural Repair*, Vol. 24, No. 7, 2010, pp. 674–679.

[16] Caria, A., Weber, C., Brotz, D., Ramos, A., Ticini, L. F., Gharabaghi, A., Braun, C., Birbaumer, N., "Chronic stroke recovery after combined BCI training and physiotherapy: A care report," *Psychophysiology*, Vol. 48, No. 4, 2011, pp. 578–582.

[17] Ang, K. K., Guan, C., Chua, K. S., Ang, B. T., Kuah, C., Wang, C., Phua, K. S., Chin, Z. Y., Zhang, H., "Clinical study of neurorehabilitationo in stroke using EEG-based motor imagery brain-computer interface with robotic feedback," *Conf Proc IEEE Eng Med Biol Soc 2010*, 2010, pp. 5549–5552.

[18] BETTER PROJECT, Brain-Neural Computer Interaction for Evaluation and Testing of Physical Therapies in Stroke Rehabilitation of Gait

Disorders. Specific Targeted Research Projects (STREP) funded by the European Commission. http://www.iai.csic.es/better/

[19] Strategic Research Program for Brain Sciences, Ministry of Education, Culture, Sports, Science and Technology http://brainprogram.mext.go.jp/

[20] Shindo, K., Kawashima, K., Ushiba, J., Ota, N., Ito, M., Ota, T., Kimura, A., Liu, M., "Effects of neurofeedback training with an electroencephalogram-based Brain Computer Interface for hand paralysis in patients with chronic stroke," *J Rehabil Med*, Vol. 43, No. 10, 2011, pp. 951–957.

[21] Ono, T., Tomita, Y., Inose, M., Ota, T., Kimura, A., Liu, M., Ushiba, J., "Multimodal sensory feedback associated with motor attempts alters BOLD responses to paralyzed hand movement in chronic stroke patients," *Brain Topogr*, in press.

[22] Mukaino, M., Ono, T., Shindo, K., Fujiwara, T., Ota, T., Kimura, A., Liu, M., Ushiba, J., "Efficacy of brain-computer interface-driven neuromuscular electrical stimulation for chronic paresis after stroke," *J Rehabil Med*, Vol. 46, No. 4, 2014, pp. 378–382.

[23] Ono, T., Shindo, K., Kawashima, K., Ota, N., Ito, M., Ota, T., Mukaino, M., Fujiwara, T., Kimura, A., Liu, M., Ushiba, J., "Brain-computer interface with somatosensory feedback improves functional recovery from severe hemiplegia due to chronic stroke," *Front Neuroeng*, Vol. 7, 2014, Article ID 19.

[24] Hashimoto, Y., Ota, T., Mukaino, M., Liu, M., Ushiba, J., "Functional recovery from chronic writer's cramp by brain-computer interface rehabilitation: a case report," *BMC Neurosci*, Vol. 15, No. 1, 2014, Article ID 103.

[25] Ushiba, J., "Application of Neurofeedback in Rehabilitation Medicine (in Japanese)," Molecular Psychiatry, Vol. 14, No. 3, 2014, pp. 164–179.

[26] Ushiba J, Kasuga S, "Recent Advances in Neuro-rehabilitation–Recovery from neurovascular disorders (in Japanese)," Neurological Therapeutics, in press.

Biographies

Dr. J. Ushiba received his Ph.D. in 2004 from School of Science and Technology, Graduate School of Keio University, Japan. He is currently Associate Professor at the Faculty of Science and Technology, Keio University. He is also an associated faculty in the Department of Rehabilitation Medicine, Keio University School of Medicine. His research interests include brain-computer interface as a neuro-rehabilitative measure and theoretical frameworks on sensorimotor nervous system and its plasticity. He has been published more than 50 peer-review English articles on proof-of-concept clinical studies and mechanism studies.

Dr. S. Kasuga received her Ph.D. in 2012 from Graduate School of Education, The University of Tokyo, Japan. She is currently a Research Associate at the Faculty of Science and Technology, Keio University, Japan. Her research interests include motor learning mechanisms, especially of upper-limb movements and its application for stroke rehabilitation. She has published 7 peer-review articles and they cover studies on the mechanism of human visuomotor learning, decision-making problems, assessment of recovery processes from spinal cord injuries, and development of an educational tool for undergraduate neuroscience courses.

3

ICT for New-Generation Prostheses

Roberto Bortoletto and Luca Tonin, Enrico Pagello,
Emanuele Menegatti

Intelligent Autonomous Systems Laboratory (IAS-Lab), Department of
Information Engineering, University of Padova, Italy
EXiMotion s.r.l., Padova, Italy
Corresponding author: Luca Tonin <luca.tonin@dei.unipd.it>

3.1 Introduction

The central nervous system (CNS) generates neural commands to activate the muscles in order to control the human body movements. Subsequent forces produced by muscles are transmitted by tendons to the skeleton to perform a motor task. Thus, muscles and tendons are the interface between the CNS and the articulated body segments. A firm understanding of the properties of the whole framework – user's intention and action's generation – is important for both scientists in order to interpret kinesiological events in the context of coordination of the body and engineers in order to design prosthetic, orthotic and functional neuromuscular stimulation systems to restore lost or impaired motor function.

From a mechanical point of view, analysis of movement plays an important role in rehabilitation of neurological and orthopaedic conditions. How mammals control the overconstrained actuation of their limb has been a research topic for decades. Researchers have tried to simplify the problem postulating that the brain uses less independent commanding signals than those necessary to control each single muscle independently. Recently, a new prospective has been proposed by Kutch and Valero-Cuevas [1]. They showed that the dimensionality reduction observed in neurophysiological studies could simply arise from the physical constraint of the system. In line with this, some studies have been recently proposed related to the estimation of joint stiffness [2] and quasi-stiffness [3–6], as an accepted parameter

Neuro-Rehabilitation with Brain Interface, 21–50.

to describe the mechanics of human limbs [7–9]. The difficulty associated with the direct measurement of important variables, including the forces generated by muscles, is one of the main limitations related to the use of experiments only. As argued by Delp et al. [10], a theoretical framework is needed, in combination with experiments, to uncover the principles that govern the coordination of muscles during normal movement, to determine how neuromuscular impairments contribute to abnormal movement and to predict the functional consequences of treatments. A dynamic simulation of movement that integrates models describing the anatomy and physiology of the elements of neuromusculoskeletal (NMS) system and the mechanics of multijoint movement provides such a framework [10].

The detection of the intention to move is the second important aspect to control new-generation prostheses for rehabilitation purposes. A brain–machine interface (BMI) or brain–computer interface (BCI) is able to recognize specific brain patterns and translate them in actual actions of external devices in order to enable a new real-time interaction between people with severe motor disabilities and the outside world. In this respect, the last years have seen an increase in sophistication of BMI-driven applications that allow direct brain communication even in completely paralysed patients and restoration of movement in paralysed limbs through the transmission of brain signals to the muscles or to external prosthetic devices [11–14]. Furthermore, evidences from several recent studies suggest that BMI technology is also mature enough to play an active role in the field of motor rehabilitation after stroke [15, 16]. In fact, BMI systems can help to promote activity-dependent brain plasticity by means of real-time analysis of brain signals, in order as to provide patients with feedback about the presence of their neural correlates of correct execution of tasks to be recovered and, if so, to trigger some therapy-relevant action. In addition, it has been proven the superior effectiveness of the BMI-based rehabilitation interventions if coupled to real and coherent movements of external actuators such as orthoses, prostheses and exoskeletons [17].

The rest of the chapter is organized as follows: in Section 3.2, the role of neuromusculoskeletal modelling in the design of new-generation prostheses is examined; in Section 3.3, the typical BMI closed loop is described; in Section 3.4, the kinematic reconstruction and goal-directed BMI approaches are treated; in Section 3.5, the available hybrid architecture is pose in relation to the role of shared-control in BMI-driven approaches. Finally, in Section 3.6, concluding discussion highlights what are the possible future trends and challenges in order to move such a new technology from simulation to laboratory, from laboratory to clinic and from clinic to patient's home.

3.2 Neuromusculoskeletal Modelling: Its Role in the Design of New-Generation Prostheses

Muscle-driven dynamic simulations rely on computational models of musculotendon dynamics. These models are commonly subdivided into two classes: cross-bridge models [18, 19] and Hill-type models [20, 21]. Although cross-bridge models have the advantage of being derived from the fundamental structure of muscle, they include many parameters that are difficult to measure and are rarely used in muscle-driven simulations involving many muscles. For this reason and due to their computational efficiency, Hill-type models are the most commonly used in muscle-driven simulations [22–29].

3.2.1 Musculotendon Models and Parameters

Biomechanical models have been used in several studies to predict the muscle forces and joint torques along with human body motion. One of the first muscle's mathematical models was proposed by Hill [30]. Gordon et al. [31] refined such a model by incorporating the dependence between changes in muscle force as function of muscle lengths and contraction speeds. Zajac extended the Hill's model introducing a muscle–tendon model [32], which is known as Hill-type muscle force model.

The general arrangement [33] for a muscle–tendon model has a muscle fibre in series with an elastic or viscoelastic tendon (Figure 3.1(A)). The muscle fibre also has a contractile element in parallel with an elastic component (Figure 3.1(B)). The Hill-type muscle model is used to estimate the force that can be generated by the contractile element of the muscle fibre, with the general form of the function given by

$$F^m(t) = f(v)\, f(l)a(t)F_0^m \tag{3.1}$$

where $F^m(t)$ represents the time-varying muscle fibre force; $f(v)$ is the normalized velocity-dependent fibre force; $f(l)$ is the normalized length-dependent fibre force; $a(t)$ is the time-varying muscle activation; and finally, F_0^m that represents the maximum isometric muscle fibre force (MIF).

The active part of muscle is due to the contractile elements. When the length of sarcomeres is optimal, the contractile elements yield a peak force, corresponding to an optimal overlap of the actin and myosin myofilaments. On the other hand, if the muscle is at a length above that optimal length, then it cannot generate as much force because there is less actin–myosin overlap which reduces the force-generating potential of the muscle.

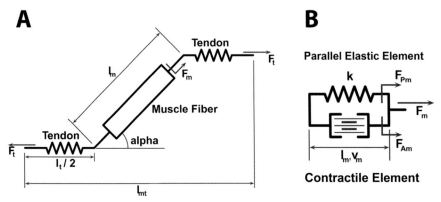

Figure 3.1   (A) Schematic of muscle–tendon unit showing muscle fibre in series with the tendon. The pennation angle, φ, of the muscle fibre relative to the tendon and that the total tendon length, l_t, is twice that of the tendon on either end of the muscle fibre, $l_t/2$. (B) Schematic of muscle fibre with the contractile element and parallel elastic component. The force produced by the contractile element, F_m, is a function of l_m (muscle fibre length) and v_m (muscle fibre contraction velocity), while the tendon force, F_t, is a function of l_t. The total muscle fibre force, F_m, is the sum of F_{Am} and F_{Pm}. *Figure modified from* [33]

It is often more helpful to consider the force–length relationship in dimensionless units, as shown in Figure 3.2(A). The passive force in the muscle is due to the elasticity of the tissue that is in parallel with the contractile

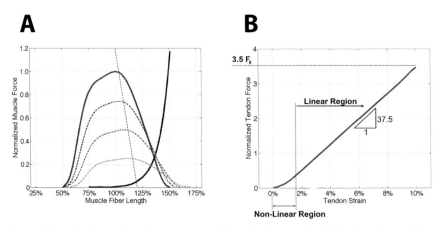

Figure 3.2   (A) Normalized force–length relationship for muscle. Thick dark lines indicate maximum activation, whereas the light thin lines are lower levels of activation. Note that the optimal fibre length is longer as the activation decreases. (B) Normalized force–length relationship for tendon. *Figure modified from* [33]

element (Figure 3.1(B)). Passive forces are very small when the muscle fibres are shorter than their optimal fibre length (OFL), l_0^m, and rise greatly thereafter.

Since the tendon is in series with the muscle, whatever force passes through the muscle must also pass through the tendon and vice versa. Tendons are passive elements that act like rubber bands. Below the tendon slack length (TSL), l_s^t, the tendon does not carry any load. However, above the TSL, it generates force proportional to the distance it is stretched (Figure 3.2(B)).

The dynamics of muscle force generation is directly related to the movement dynamics of the skeletal system. Thus, modelling muscle dynamics is important to fully understand the control of movement in humans. Two major, freely available, implementations of the Hill-type muscle model are widely used: *Thelen 2003 Muscle Model*[1] [34] and *Millard 2012 Muscle Model*[2] [35].

3.2.2 Musculoskeletal Kinematics

Once the muscle–tendon force is computed, it is important to compute the corresponding contribution to joint moment. This requires the knowledge of the muscle's moment arm, r, which can be shown to be a function of the muscle's length [36]. To compute both the length and the moment arm for a musculotendon unit, a musculoskeletal model is required. Musculotendon kinematics estimations can be produced by software that models the geometry of the bones, the complex relationships associated with joint kinematics and the musculotendon paths wrapping around points and surfaces [10]. This is based on obstacle detection and may cause discontinuities in the predicted musculotendon kinematics [37]. On the other hand, it is desirable that musculotendon kinematics equations are continuously differentiable to enable the computation of analytical Jacobians for the forward simulation of the musculoskeletal system [22]. Recently, a robust and computationally inexpensive method to estimate the lengths and three-dimensional moment arms for a large number of musculotendon actuators of the human lower limb has been presented [38]. It is based on the use of a musculoskeletal model of the lower extremity and of a multidimensional spline function used to fit the length of each musculotendon actuator to different lower limb generalized coordinates (joint angles).

---

[1] *Thelen 2003 Muscle Model*: http://simtk-confluence.stanford.edu:8080/display/OpenSim/Thelen+2003+Muscle+Model
[2] *Millard 2012 Muscle Model*: http://simtk-confluence.stanford.edu:8080/display/OpenSim/Millard+2012+Muscle+Models

3.2.3 Neuromuscular Control of Musculoskeletal Systems

Muscle redundancy, having more muscles than mechanical degrees of freedom (DOFs), has long been a central problem in biomechanics and neural control [39, 40]. The issue concerns how the CNS selects muscle coordination patterns from a theoretically infinite set of possibilities [41]. Scientists suggest that humans centrally control their movements in feedforward manner and locally in feedback way [42], also known as open loop and closed loop, respectively [43].

Generally speaking, actions can be considered as operational modules in which descending motor patterns are produced together with the expectation of the (multimodal) sensory consequences. Mounting evidence accumulated in the last decades from different directions and points of view, such as the equilibrium point hypothesis [44–47], mirror neurons system [48], motor imagery [49–51], motor resonance [52] and embodied cognition [53, 54], suggests that in order to understand the neural control of movement, the observation, and analysis of overt movements, is just the tip of the iceberg because what really matters is the large computational basis shared by "*Action Production*", "*Action Observation*", "*Action Reasoning*", and "*Action Learning*".

Muscle coordination studies repeatedly show low dimensionality of muscle activations for a wide variety of motor tasks. The basis vectors of this low-dimensional subspace, termed muscle synergies, are hypothesized to reflect neurally established functional muscle groupings that simplify body control [55–57]. However, the muscle synergy hypothesis has been notoriously difficult to prove or falsify [58]. An alternative explanation of how muscle synergies could be observed has been provided by Kutch and Valero-Cuevas [1]. They used cadaveric experiments and computational models to perform a crucial thought experiment and develop an alternative explanation of how muscle synergies could be observed without the nervous system having controlled muscles in groups. They first showed that the biomechanics of the limb constrains musculotendon length changes to a low-dimensional subspace across all possible movement directions. Then, a modest assumption that each muscle is independently instructed to resist length change leads to the result that electromyography synergies will arise without the need to conclude that they are a product of neural coupling among muscles. Finally, they showed that there are dimensionality-reducing constraints in the isometric production of force in a variety of directions, but that these constraints are more easily controlled for, suggesting new experimental directions.

3.2.4 Patient-Specific Neuromusculoskeletal Modelling

Every patient is different and possesses unique anatomical, neurological and functional characteristics that may significantly affect the optimal treatment of the patient. Personalized computational models of the NMS system can facilitate prediction of patient-specific functional outcomes for different treatment designs and provide useful information for clinicians. Personalized models may reduce the likelihood that different clinicians will plan different treatments given the same patient data [59]. Depending on the intended clinical application, a personalized NMS model might account for patient-specific anatomical (e.g. skeletal structure and muscle lines of action), physiological (e.g. muscle force-generating properties) and/or neurological (e.g. constraints on achievable muscle excitation patterns) characteristics, all within the context of a multibody dynamic model [59]. Among the numerous tools and theoretical frameworks available, we focus here on some of the major open-source and proprietary solutions.

OpenSim[3] is an open-source platform for modelling, simulating and analysing the neuromusculoskeletal system [10]. It includes low-level computational tools that are invoked by an application. A graphical user interface provides access to key functionality. OpenSim is being developed and maintained on Simtk.org[4] by a growing group of participants. Simtk.org serves as a public repository for data, models and computational tools related to physics-based simulation of biological structures.

CEINMS[5] (*Calibrated EMG-Informed Neuromusculoskeletal Modelling Toolbox*) permits the simulation of all the transformations that take place from the onset of muscle excitation to the generation of force. Experimentally recorded electromyography (EMG) signals and three-dimensional joint angles can be used to determine the neural drive and the instantaneous kinematics for the multiple musculotendon units being modelled.

MSMS[6], developed by *Medical Device Development Facility – University of Southern California*, is a software application for modelling and simulation of neural prostheses systems. It can be used to model and simulate human and prosthetic limbs and the task environment they operate in. The simulations can be executed in a stand-alone computer to develop and test neural control

---

[3] *OpenSim* Project home page: https://simtk.org/home/opensim
[4] *Simtk.org* home page: https://simtk.org/xml/index.xml
[5] *CEINMS* Project home page: https://simtk.org/home/ceinms
[6] *MSMS* Project home page: http://mddf.usc.edu:85/?page_id=94

systems or in a virtual reality environment where the human or animal subject can interact with and therefore affect the behaviour of the simulated limb.

SIMM[7] (*Software for Interactive Musculoskeletal Modelling*) is a powerful proprietary toolkit that facilitates the modelling, animation and analysis of 3D musculoskeletal systems. In SIMM, a musculoskeletal model consists of representations of bones, muscles, ligaments and other structures. Muscles span the joints and develop force, thus generating moments about the joints. SIMM enables an analysis of a musculoskeletal model by calculating the joint moments that each muscle can generate at anybody's position. By manipulating a model using the graphical interface, the user can quickly explore the effects of changing musculoskeletal geometry and other model parameters on muscle forces and joint moments.

MSC Software's Adams[8] is a proprietary tool used for simultaneous prediction of knee loading and muscle forces during human movement. The multibody modelling software simulates the articulations of cartilage-covered bone, the forces of ligaments crossing the knee, as well as the dynamic motion and loading of knee menisci. The patient-specific knee models are placed within body-level musculoskeletal models that include prediction of muscle forces as well as contact between the foot and ground. Loading on knee structures, such as ligaments and cartilage, is then predicted during muscle force-driven simulations of walking or other activities.

3.3 Catching the Intention to Move: Brain–Machine Interfaces

It was 1929 when Hans Berger performed the first human electroencephalogram (EEG) recording and proposed the pioneering idea about the possibility of detecting and reading "thoughts" from the EEG activity. Even more important, he approached the dichotomous question "human brain" and "human mind" by considering a direct and physiological relationship between them [60]. However, in the following decades, the EEG has been used mainly in the evaluation and investigation of neurological disorders. The most common approaches are reviewed at [61]. Only recent research and technology development have made it possible to start deciphering intents and task-related – actual or imagined – brain patterns by the EEG. A final opportunity has come

---

[7]*SIMM* home page: http://www.musculographics.com/html/products/SIMM.html

[8]*Adams* home page: http://www.mscsoftware.com/it/academic-case-studies/msc-softwares-adams-used-musculoskeletal-models-simultaneously-predict-knee

to create "*a communication system that does not depend on the brain's normal output of peripheral nerves or muscles*" [62] in order to enhance the quality of life of those who – due to different neurological disorders – are not able to control their own body anymore. This is a widely accepted definition of brain–machine interface. Several brain signals (and modalities) have been investigated to control such an interface, and over the years, this definition has been changed and extended in order to include last findings and possible new applications.

Among the different mental tasks and brain signals that have been taken into consideration, most BMI systems share the same architecture and technological components. Figure 3.3 illustrates the so-called *BMI closed loop* defined for the first time by Wolpaw [62]. The main components may be divided into signal acquisition, feature extraction, classification and action generation. Signals are acquired from the user's brain, while he/she is performing a predefined mental task (e.g. imagination of hand movement) and those specific features that are supposed to better encode user's intention are extracted. Then, features are classified and translated into an output signal (e.g. posterior probabilities associated with user's intent) by means of machine

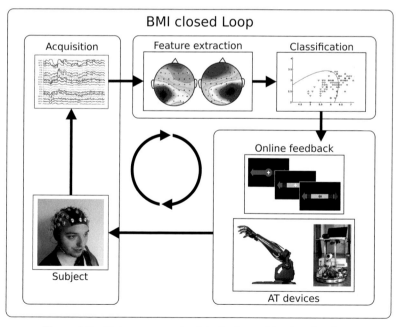

Figure 3.3 Main components of the brain-machine interface loop

learning algorithms. Finally, the output of the process is used as a control signal for external devices.

Besides the architecture of a BMI system, the key characteristic is to provide a real-time online feedback in order to maintain the direct connection between user's intent and the generated action. On the one side, the feasibility to recognize specific mental task and on the other, the possibility to learn how to modulate brain patterns by means of operant conditioning – or directly by neurofeedback – are the two fundamental pillars of a brain–machine interface system. The strict interaction between these two aspects implements the *closed-loop* approach where the BMI user may continuously self-regulate his brain activity according to the provided real-time online feedback. The user can learn the best strategy to operate the BMI.

BMIs can be invasive or non-invasive according to the techniques that have been used to record brain signals and, consequently, user's intents. An invasive BMI relies on the activation of single neurons or population of neurons recorded by means of a multiunit array implanted within the cerebral cortex or its surfaces. These systems directly record the neuronal firing rates or the local field potential (LFP) [63–65]. Regardless of the several advantages in terms of quality of signals, signal-to-noise ratio and spatial resolution, invasive BMIs suffer from substantial drawbacks related to technical difficulties and clinical risks of a surgical intervention. The main disadvantages of an invasive approach range from ethical to more technical issues: from the user's safety (i.e. risk of infection and other damages to the brain) to the quality of signals due to the formation of scar tissue and the deterioration of the implanted electrodes. Therefore, in the literatures, these systems have been mainly investigated in primates and only few attempts have been effectively reported with human subjects.

Contrariwise, non-invasive BMIs decode user's intents from activity recorded at the scalp level. In the literature, several non-invasive acquisition techniques have been investigated based on the modulation of both the electrical activity of the brain (i.e. EEG and magnetoencephalogram (MEG)-based BMI) and the blood oxygen level (i.e. functional magnetic resonance imaging (fMRI), near-infrared spectroscopy (NIRS)) [66–71]. The main advantage of these acquisition techniques is that they do not require any kind of surgical intervention. In addition, in the case of EEG- and NIRS-based BMIs, the system is portable and low cost and it may be set up in a short period both in a clinical environment and at a patient's home. The low signal-to-noise ratio and the poor spatial or temporal resolution are the intrinsic disadvantages of such methods. Nonetheless, non-invasive BMIs are extensively utilized in studies involving human subjects, and several approaches (i.e. in terms of signal processing or control theory) have been developed in order to face these limitations.

In this context, one of the main challenges in the research field is to employ BMI to catch the user's intention of moving and to naturally control new-generation robotic prostheses by bypassing conventional – and easily damaged – activation channels.

3.4 Kinematic Reconstruction and Goal-Directed BMI Approaches

The last years have seen several demonstrations of BMI systems for controlling mobile devices (e.g. semi-autonomous wheelchairs and telepresence robots [72–74, 14, 13, 75]) and for directly operating neuroprosthesis devices (e.g. robotic arms, upper/lower limb exoskeletons [11, 12, 76]).

New-generation prostheses can be BMI-controlled based on two different modalities: by kinematic reconstruction and by goal-directed approach [77]. In the first case, the exact device's motion parameters are extracted from user's brain patterns, while the user has continuous and direct real-time control of the device (e.g. a robotic arm) as an extension of his body. The most intriguing results in kinematic reconstruction have been reached by means of invasive BMI systems [11, 12].

In particular, Carmena et al. [11] demonstrated the possibility to train monkeys to brain-control a robotic arm without any muscular activity. They employed neural firing rates recorded by multiple arrays implanted at the level of primary, supplementary and premotor cortex in order to predict and estimate position, velocity and force during reaching and grasping tasks. Results showed high control performances and that direct neural control of prosthetic devices is possible.

Recently, Hochberg et al. achieved similar results even in the case of human subjects [12]. In this study, two participants with tetraplegia and anarthria, as a result of brainstem stroke, were able to neural operate a robotic in undercontrol conditions. Even more interesting for daily life, a participant could use the robotic arm to grip a glass and drink it by means of the mind-controlled robotic arm.

In parallel with the invasive approach, several groups (especially in Europe and in Asia) started investigating the feasibility and the suitability of non-invasive EEG-based BMI systems for the kinematic control of robotic devices and neuroprostheses. However, the possibility to decode detailed movement information directly from EEG is still up for debate. In fact, if, on the one hand, evidences from the literature showed that low-frequency EEG encodes information of hand trajectory both in 2D and in 3D [78–81], on the other, the effectiveness of this method is still to be proven [82, 83].

The second BMI control modality relies on a goal-directed approach to operate the device. In this case, the BMI provides high-level commands related to desired outcome (e.g. move left, move right, reach the selected target) and the system is in charge of generating the overall kinematic necessary to achieve this outcome. From the user's point of view, goal-directed approach is much less demanding in terms of workload because most of the complexity is charged on the robotic device.

The goal-directed approach is mainly implemented by non-invasive EEG BMI systems where users can control the device by modulating different kinds of brain signals. We will focus mainly on those BMIs related to the self-paced, voluntary modulation of brain activity and to a direct link to the intention of moving. For a full review of all BMI types (e.g. BMIs based on exogenous stimulation, aka steady-state visual-evoked potential or P300 BMIs), please refer to [84].

The first BMIs were adopting the so-called slow cortical potential (SCP) to allow people suffering from amyotrophic lateral sclerosis (ALS) to communicate with the external world [85–87]. SCPs are distinguished by a contingent negativity (or positivity) in the central electrodes over relatively long time windows (0.5–10 s). They are associated with actual movements or movement preparation and – in general – with functions involving cortical activation. Long-term studies demonstrated the possibility to learn how to modulate these particular brain patterns and the feasibility to use them as a binary control signal for BMI-driven devices (i.e. text speller [88]). Furthermore, recent findings suggest that single-trial classification of the anticipatory behaviour in SCP is possible [89].

Following the idea of self-regulation of brain rhythms, researchers have started developing a new branch of BMIs based on sensory motor rhythm (SMR). It has been proved that during actual or imagined movements (e.g. imagination of right-/left-hand movement), an event-related desynchronization (ERD) and event-related synchronization (ERS) appear over the sensorimotor cortex in the so-called μ-rhythm (10–12 Hz) [90, 91]. For instance, SMR-BMIs rely on this well-defined brain activity (in space and frequency) to control a cursor on the screen and to send predefined commands to different kinds of assistive technology (AT) devices [92, 73, 93, 13, 75, 72].

It is worth noting that the user can see in real time the visual representation of the system output and, thus, can learn the best mental strategy to send the desired command. At the same time, the system needs to be updated over time to the current user's brain patterns. In this sense, SMR-BMIs fully implement the mutual learning interaction between user and computer, eventually closing the BMI loop.

In parallel with the aforementioned EEG-based BMI, several studies with MEG, fMRI and NIRS demonstrated the attainability of using alternative neurophysiological process as a control signal for BMIs [66–71]. Each system carries its own advantages and disadvantages (e.g. high spatial vs. low temporal resolution in the case of blood-oxygen-level-dependent contrast (BOLD) signals for fMRI). Nevertheless, the main drawbacks of these systems are still the cost and the transportability – even more so if the final target is a daily operation by disabled people.

At the current state, non-invasive BMIs (e.g. EEG, NIRS based) can be resorted in order to detect user intentions and to provide external devices with high-level control commands. They compensate for the lower precision and control resolution with respect to invasive systems, with lower cost, more safety and suitability in clinics or in daily-life applications. Furthermore, several techniques have been developed in order to extend and improve the operations of non-invasive BMI-driven devices by allowing the user direct mind control in natural and complex situations.

3.5 The Hybrid Architecture and the Role of Shared-Control in BMI-Driven Devices

As already mentioned, the control of devices by means of an EEG-based BMI suffers from intrinsic limitations such as the number of available commands, the signal reliability and the level of effort for the user. Recently, a new and promising approach has been proposed in order to overtake these drawbacks: a hybrid BCI system (hBCI).

As a general definition, an hBCI is a combination of different signals including at least one brain channel (e.g. EEG plus EMG biosignals, see Figure 3.4) [94]. Thus, a user suffering from the progressive loss of muscular activity might control the movement of a robotic arm by means of his residual motor functionalities, while he uses his motor imagination only to perform the action of grasping. Even more so, an hBCI-driven device might autonomously switch from a manual control modality to full BMI-based operations when muscular fatigue is detected (i.e. at the end of the day). Furthermore, all the

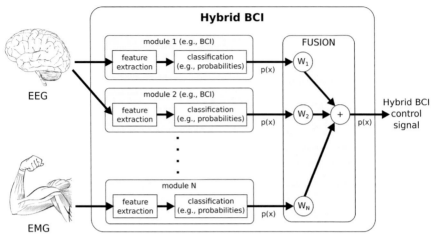

Figure 3.4   Illustration of the principles of hBCI architecture (*Figure modified from* [94])

input channels can be weighted and fused together in order to give a more reliable and robust control signal.

Currently, BMI research is moving fast towards a hybrid approach to solve the issue of operating complex devices in natural and daily-life situations. Several examples that adopt such a design have been reported, in the literature, to control software applications [95–97, 92]. External navigation devices [74] and even new-generation neuroprostheses [98, 97]. The achieved results demonstrated that an EEG BMI-driven device is not only attainable but can also reach high levels of performance and usability for disabled users.

In the context of hybrid architecture for BMI, one of the most promising techniques for operating complex BMI-driven devices is the shared-control approach. The role of shared-control is to contextualize high-level commands in the current situation and thus to allow the device to perform a wide range of behaviours with minimum user effort (Figure 3.5(A)). The concept of shared-control comes from the human–computer (robot) interaction field and it is based on the so-called H-Metaphor. The H-Metaphor (or Horse-Metaphor) was proposed by Flemish et al. in 2003 [99] and implements a new design for semi-autonomous vehicles as simple as effective.

Imagine you are riding a horse in a wooded park: you need to focus your attention only on the final destination of your journey, or you can even just enjoy the scenery, without caring about the low-level navigation details. In fact, you are confident that the horse is intelligent enough to avoid possible obstacles and to follow the correct path. At the same time, you have the

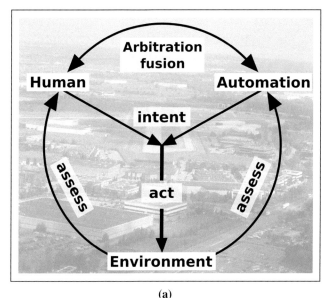

(a)

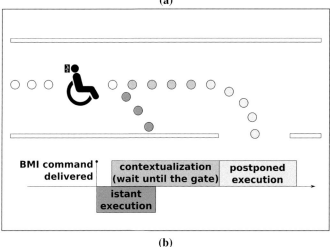

(b)

Figure 3.5 (A) Shared control between cooperative agents (*Figure modified from* [102]). (B) Typical situation where user's command has to be contextualized by shared-control

full control of your "vehicle": by means of a simple movement of the reins, you can make it turn, stop or move forward. This idea perfectly matches the control limitation of a non-invasive BMI system that – as already mentioned – conveys few and high-level commands associated with specific mental tasks performed by the user (e.g. imagination of right- or left-hand movement). In this framework, a BMI-driven device, such as a telepresence robot or a semi-autonomous wheelchair or even a lower limb exoskeleton, may perform complex behaviours (e.g. moving in a real and crowded environment, docking a table, entering a door) by means of a few user's commands. The key point of such a system is the capability of contextualizing the same command according to different situations in order to perform the intended action. A typical example refers to the "*gate problem*" (Figure 3.5(B)). Imagine the user is mentally driving his wheelchair along a corridor with several doors. The user wants to enter in the next door, and thus, he mentally sends the "turn" command to the wheelchair. However, if the wheelchair implemented the intended action as soon as the command is received, it would crash into the wall. The shared-control should be intelligent enough to understand the user's intent (entering the door) and postpone the turn until the door is reached. Contrariwise, if the wheelchair was in an open space (i.e. where there are no environmental constraints), the shared-control should make the user free to turn at any time he wants.

Over the last years, shared-control has been widely utilized in BMI-based applications. For instance, Millán's group demonstrated the possibility for disabled users to remotely mentally drive a telepresence robot [13, 75, 74] and the feasibility of a BMI-driven semi-autonomous wheelchair [100]. In both cases, results highlighted that the role of shared-control is twofold: on the one hand, it helps disabled users to accomplish the navigation task; on the other, it reduces the workload and allows them to reach similar performances as healthy subjects (i.e. time, number of commands to complete the task).

3.6 Future Challenges for New-Generation Prostheses

The successful control of neuroprostheses may be achieved only if we are able to correctly detect user's motor intents and to translate them in coherent, natural and ecological actions of the new-generation ICT devices. Challenges are open from both theoretical and technological points of view in order to demonstrate that neuroprostheses systems are mature enough to be used in real- and daily-life situations.

Looking at the characterization of the patient's movement biomechanics, a major challenge is related to the fact that muscle–tendon forces, joint contact forces and several other physiological parameters are currently not measurable *in vivo* with non-invasive devices, during motion. Furthermore, often, the measurement system alters the movement of the patient falsifying the measurement itself. Computational modelling of the musculoskeletal system is the only practicable method that can provide an approach to analyse loading of muscle and joint. The development of new modelling methods and numerical simulation algorithms, which are computationally efficient, is increasingly raising the interest in musculoskeletal modelling and simulation among the biomechanical and medical communities.

On the other hand, considering the modelling of the interaction between patient and orthotics, two critical tasks in process of using personalized models are calibration and validation [59]. Since generic models are constructed from detailed anatomic measurements taken on cadaver specimens, a subject-specific model calibration is needed. Four calibration steps of the proposed model that should be performed in whole or in part to transform a generic model into a personalized model include geometric calibration, kinematic calibration, kinetic calibration and neurologic calibration. Validation of clinical predictions is the other major challenge faced by personalized models that will ultimately require randomized controlled trials, where outcomes are compared between patients whose treatments were planned with a personalized model and those whose treatments were not.

The fundamental requirement for the control of BMI-driven neuroprostheses is to improve the accessibility, reliability and robustness of such systems in decoding user's intentions. In this respect, user's brain patterns have to be acquired by means of non-invasive techniques while ensuring the quality of the signals. Improvements in the acquisition devices (e.g. amplifier, electrodes) as well as advances in the decoding algorithms are strictly required in order to bring BMI system "out-of-the-laboratory". As previously mentioned, researchers are currently investigating novel approaches (i.e. hybrid BMI systems, shared-control theory) to fulfil this ambitious goal. Next challenge is to implement and integrate them in the new generation of prostheses: on the one side, a hybrid control of the prostheses that fuses together different kinds of physiological signals (e.g. EEG, EMG, residual muscular functionality) in order to enhance the quality and the naturalness of the control. On the other side, the goal is to improve the human–robot interaction by means of intelligent prostheses able to gather information from the environment, to take autonomous decisions, to adapt its behaviour to the musculoskeletal parameters of the user and eventually, to provide him with an haptic and coherent feedback. In this context, the strict coupling between human intention, brain–machine interface and intelligent device will make the prostheses no longer an external actuator, but an actual extension of the user's body.

References

[1] J. Kutch e F. Valero-Cuevas, «Challenges and New Approaches to Proving the Existence of Muscle Synergies of Neural Origin,» *PLoS Comput Biol,* vol. 8, n. 5, p. online, 2012.

[2] R. Bortoletto, E. Pagello e D. Piovesan, «Lower Limb Stiffness Estimation during Running: The Effect of Using Kinematic Constraints in Muscle Force Optimization Algorithms,» in *Lecture Notes in Computer Science: Simulation, Modeling, and Programming for Autonomous Robots (SIMPAR),* Bergamo, 2014.

[3] R. Bortoletto, E. Pagello e D. Piovesan, «A comparison between multi-joint stiffness and quasi-stiffness in humanoid locomotion,» in *Preprint submitted to 2015 IEEE Int. Conf. on Robotics and Automation (ICRA). UNDER REVIEW*, Seattle, WA, USA, 2015.

[4] E. Rouse, R. Gregg, L. Hargrove e J. Sensinger, «The Difference Between Stiffness and Quasi-Stiffness in the Context of Biomechanical Modeling,» *IEEE Transactions on Biomechanical Modeling, Biomedical Engineering*, vol. 60, n. 2, pp. 562–568, 2013.

[5] K. Shamaei, G. Sawicki e A. Dollar, «Estimation of Quasi-Stiffness of the Human Knee in the Stance Phase of Walking,» *PLoS ONE*, vol. 8, n. 3, p. online, 2013.

[6] K. Shamaei, G. Sawicki e A. Dollar, «Estimation of Quasi-Stiffness and Propulsive Work of the Human Ankle in the Stance Phase of Walking,» *PLoS ONE*, vol. 8, n. 3, p. online, 2013.

[7] F. Mussa-Ivaldi, N. Hogan e E. Bizzi, «Neural, mechanical, and geometric factors subserving arm posture in humans,» *Journal of Neuroscience*, vol. 5, n. 10, pp. 2732–2743, 1985.

[8] D. Piovesan, P. Morasso, P. Giannoni e M. Casadio, «Arm Stiffness During Assisted Movement After Stroke: The Influence of Visual Feedback and Training,» *IEEE Transactions on Neural Systems and Rehabilitation Engineering*, vol. 21, n. 3, pp. 454–465, 2013.

[9] D. Piovesan, A. Pierobon, P. Dizio e J. Lackner, «Experimental measure of arm stiffness during single reaching movements with a time-frequency analysis,» *Journal of Neurophysiology*, vol. 110, n. 10, pp. 2484–2496, 2013.

[10] S. Delp, F. Anderson, A. Arnold, P. Loan, A. Habib, C. John, E. Guendelman e D. Thelen, «OpenSim: open-source software to create and analyze dynamic simulations of movement,» *IEEE Transactions on Biomedical Engineering*, vol. 54, n. 11, pp. 1940–1950, 2007.

[11] J. M. Carmena, M. A. Lebedev, R. E. Crist, J. E. O'Doherty, D. M. Santucci, D. F. Dimitrov, P. G. Patil, C. S. Henriquez e M. A. Nicolelis, «Learning to control a brain-machine interface for reaching and grasping by primates,» *PLoS Biology*, vol. 1, n. 2, p. E42, 2003.

[12] L. R. Hochberg, D. Bacher, B. Jarosiewicz, N. Y. Masse, J. D. Simeral, J. Vogel, S. Haddadin, J. Liu, S. S. Cash, P. van der Smagt e J. P. Donoghue, «Reach and grasp by people with tetraplegia using a neurally controlled robotic arm,» *Nature*, vol. 485, n. 7398, pp. 372–5, 2012.

[13] L. Tonin, R. Leeb, M. Tavella, S. Perdikis e J. d. R. Millán, «The role of shared-control in BCI-based telepresence,» in *29th IEEE International Conference on Systems, Man and Cybernetics*, 2010.

[14] L. Tonin, T. Carlson, R. Leeb e J. d. R. Millán, «Brain-controlled telepresence robot by motor-disabled people,» in *33rd Annual International Conference of the IEEE Engineering in Medicine and Biology Society*, 2011.

[15] J. J. Daly e J. R. Wolpaw, «Brain-computer interfaces in neurological rehabilitation,» *Lancet Neurology*, vol. 7, n. 11, pp. 1032–43, 2008.

[16] M. Grosse-Wentrup, D. Mattia e K. Oweiss, «Using brain-computer interfaces to induce neural plasticity and restore function,» *Journal of Neural Engineering*, vol. 8, n. 2, p. 025004, 2011.

[17] A. Ramos-Murguialday, D. Broetz, M. Rea, L. Läer, O. Yilmaz, F. L. Brasil, G. Liberati, M. R. Curado, E. Garcia-Cossio, A. Vyziotis, W. Cho, M. Agostini, E. Soares, S. Soekadar, A. Caria, L. G. Cohen e N. Birbaumer, «Brain-machine interface in chronic stroke rehabilitation: a controlled study,» *Annals of Neurology*, vol. 74, n. 1, pp. 100–8, 2013.

[18] E. Eisenberg, T. Hill e Y. Chen, «Cross-bridge model of muscle contraction,» *Biophisical Journal*, vol. 29, n. 2, pp. 195–227, 1980.

[19] G. Zahalak e S. Ma, «Muscle activation and contraction: constitutive relations based directly on cross-bridge kinetics,» *Journal of Biomechanical Engineering*, vol. 112, n. 1, pp. 52–62, 1990.

[20] M. Epstein e W. Herzog, Theoretical Models of Skeletal Muscle: Biological and Mathematical Considerations, New York: Wiley; 1st edition, 1998.

[21] J. Winters e L. Stark, «Muscle models: What is gained and what is lost by varying model complexity,» *Biological Cybernetics*, vol. 55, n. 6, pp. 403–420, 1987.

[22] M. Ackermann e A. van den Bogert, «Optimality principles for model-based prediction of human gait,» *Journal of Biomechanics*, vol. 43, n. 6, pp. 1055–1060, 2010.

[23] F. Anderson e M. Pandy, «Static and dynamic optimization solutions for gait are practically equivalent,» *Journal of Biomechanics*, vol. 34, n. 2, pp. 153–161, 2001.

[24] E. Arnold e S. Delp, «Fibre operating lengths of human lower limb muscles during walking,» *Philosophical transactions of the Royal Society of London. Series B, Biological Sciences*, vol. 366, n. 1570, pp. 1530–1539, 2011.

[25] S. Hamner, A. Seth e S. Delp, «Muscle contributions to propulsion and support during running,» *Journal of Biomechanics*, vol. 43, n. 14, pp. 2709–2716, 2010.

[26] M. Liu, F. Anderson, M. Pandy e S. Delp, «Muscles that support the body also modulate forward progression during walking,» *Journal of Biomechanics*, vol. 39, n. 14, pp. 2623–2630, 2006.

[27] K. Steele, A. Seth, J. Hicks, M. Schwartz e S. Delp, «Muscle contributions to support and progression during single-limb stance in crouch gait,» *Journal of Biomechanics*, vol. 43, n. 11, pp. 2099–2105, 2010.

[28] F. Zajac, R. Neptune e S. Kautz, «Biomechanics and muscle coordination of human walking. Part I: introduction to concepts, power transfer, dynamics and simulations,» *Gait & Posture*, vol. 16, n. 3, pp. 215–232, 2002.

[29] F. Zajac, R. Neptune e S. Kautz, «Biomechanics and muscle coordination of human walking: part II: lessons from dynamical simulations and clinical implications,» *Gait & Posture*, vol. 17, n. 1, pp. 1–17, 2003.

[30] A. Hill, «The heat of shortening and the dynamic constants of muscle,» *Royal Society of London. Series B, Biological Sciences*, vol. 126, n. 843, pp. 136–195, 1938.

[31] A. Gordon, A. Huxley e F. Juliank, «The variation in isometric tension with sarcomere length in vertebrate muscle fibres,» *The Journal of Physiology*, vol. 184, n. 1, pp. 170–192, 1966.

[32] F. Zajac, «Muscle and tendon: properties, models, scaling, and application to biomechanics and motor control,» *Critical reviews in biomedical engineering*, vol. 17, n. 4, pp. 359–411, 1989.

[33] T. Buchanan, D. Lloyd, K. Manal e T. Besier, «Neuromusculoskeletal Modeling: Estimation of Muscle Forces and Joint Moments and

Movements From Measurements of Neural Command,» *Journal of Applied Biomechanics*, vol. 20, n. 4, pp. 367–395, 2004.

[34] D. Thelen, «Adjustment of muscle mechanics model parameters to simulate dynamic contraction in older adults,» *ASME Journal of Biomechanical Engineering*, vol. 125, n. 1, pp. 70–77, 2003.

[35] M. Millard, T. Uchida, A. Seth e S. Delp, «Flexing computational muscle: modeling and simulation of musculotendon dynamics,» *ASME Journal of Biomechanical Engineering*, vol. 135, n. 2, p. 11, 2013.

[36] L. Menegaldo, A. de Toledo Fleury e H. Weber, «Moment arms and musculotendon lengths estimation for a three-dimensional lower-limb model,» *Journal of Biomechanics*, vol. 37, n. 9, pp. 1447–1453, 2004.

[37] B. Garner e M. Pandy, «The Obstacle-Set Method for Representing Muscle Paths in Musculoskeletal Models,» *Computer Methods in biomechanics and biomedical engineering*, vol. 3, n. 1, pp. 1–30, 2000.

[38] M. Sartori, M. Reggiani, A. van den Bogert e D. Lloyd, «Estimation of musculotendon kinematics in large musculoskeletal models using multidimensional B-splines,» *Journal of Biomechanics*, vol. 45, n. 3, pp. 595–601, 2012.

[39] J. Kutch e F. Valero-Cuevas, «Muscle redundancy does not imply robustness to muscle dysfunction,» *Journal of Biomechanics*, vol. 44, n. 7, pp. 1264–1270, 2011.

[40] M. Latash, «The bliss (not the problem) of motor abundance (not redundancy),» *Experimental brain research*, vol. 217, n. 1, pp. 1–5, 2012.

[41] N. Bernstein, The Co-ordination and Regulation of Movements, Oxford: Pergamon Press, 1967.

[42] M. Hinder e T. Milner, «The case for an internal dynamics model versus equilibrium point control in human movement,» *The Journal of Physiology*, vol. 549, n. 3, pp. 953–963, 2003.

[43] J. Collins e C. De Luca, «Open-loop and closed-loop control of posture: a random-walk analysis of center-of-pressure trajectories,» *Experimental brain research. Experimentelle Hirnforschung, Experimentation cerebrale*, vol. 95, n. 2, pp. 308–318, 1993.

[44] D. Asatryan e A. Feldman, «Functional tuning of the nervous system with control of movements or maintenance of a steady posture,» *Biophysics*, vol. 11, n. -, pp. 925–935, 1965.

[45] E. Bizzi, A. Polit e P. Morasso, «Mechanisms underlying achieve-ment of final position,» *Journal of Neurophysiology*, vol. 39, n. -, pp. 435–444, 1976.

[46] E. Bizzi, N. Hogan, F. Mussa Ivaldi e S. Giszter, «Does the nervous system use equilibrium-point control to guide single and multiple joint movements?,» *Behav. Brain Sci.*, vol. 15, n. -, p. 603, 1992.

[47] A. Feldman e A. Levin, «The origin and use of positional frames of reference in motor control,» *Behav. Brain Sci.*, vol. 18, n. 4, pp. 723–744, 1995.

[48] G. Di Pellegrino, L. Fadiga, L. Fogassi, V. Gallese e G. Rizzolatti, «Understanding motor events: a neurophysiological study,» *Exp. Brain Res.*, vol. 91, n. -, pp. 176–180, 1992.

[49] S. Grafton, «Embodied cognition and the simulation of action to understand others,» *Ann. N. Y. Acad. Sci.*, vol. 117, n. -, pp. 1156–1197, 2009.

[50] C. Kranczioch, S. Mathews, J. Dean e A. Sterr, «On the equivalence of executed and imagined movements,» *Hum. Brain Mapp.*, vol. 30, n. 10, pp. 3275–3286, 2009.

[51] J. Munzert, B. Lorey e K. Zentgraf, «Cognitive motor processes: the role of motor imagery in the study of motor representations,» *Brain Res. Rev*, vol. 60, n. 2, pp. 306–326, 2009.

[52] P. Borroni, A. Gorini, G. Riva, S. Bouchard e G. Gabriella Cerri, «Mirroring avatars: dissociation of action and intention in human motor resonance,» *Eur. J. Neurosci.*, vol. 34, n. 4, pp. 662–669, 2011.

[53] V. Gallese e C. Sinigaglia, «What is so special with embodied simulation?,» *Trends Cogn. Sci.*, vol. 15, n. 11, pp. 512–519, 2011.

[54] V. Sevdalis e P. Keller, «Captured by motion: dance, action understand-ing, and social cognition,» *Brain Cogn.*, vol. 77, n. 2, pp. 231–236, 2011.

[55] A. d'Avella, P. Saltiel e E. Bizzi, «Combinations of muscle synergies in the construction of a natural motor behavior,» *Nature Neuroscience*, vol. 6, n. 3, pp. 300–308, 2003.

[56] L. Ting e J. Macpherson, «A limited set of muscle synergies for force control during a postural task,» *Journal of Neurophysiology*, vol. 93, n. 1, pp. 609–613, 2005.

[57] M. Tresch, P. Saltiel e E. Bizzi, «The construction of movement by the spinal cord,» *Nature Neuroscience*, vol. 2, n. 2, pp. 162–167, 1999.

[58] M. Tresch e A. Jarc, «The case for and against muscle synergies,» *Curr Opin Neurobiol*, vol. 19, n. 6, pp. 601–607, 2009.

[59] B. Fregly, M. Boninger e D. Reinkensmeyer, «Personalized neuromusculoskeletal modeling to improve treatment of mobility impairments:
a perspective from European research sites,» *Journal of Neuroengineering and Rehabilitation,* vol. 9, n. 1, p. 18, 2012.

[60] P. Gloor, «Berger lecture. Is Berger's dream coming true?,» *Electroen-
cephalgraphy and Clinical Neurophysiology,* vol. 90, n. 4, pp. 253–66,
1994.

[61] B. M. Sterman, «The brain response interface: Communication
through visual-induced electrical brain responses,» *Journal of Micro-
computer Applications,* vol. 15, n. 1, pp. 31–45, 1992.

[62] J. R. Wolpaw, N. Birbaumer, W. J. Heetderks, D. J. McFarland, P. H.
Peckham, G. Schalk, E. Donchin, L. A. Quatrano, C. J. Robinson e
T. M. Vaughan, «Brain-computer interface technology: A review of
the first international meeting,» *IEEE Transactions on Rehabilitation
Engineering,* vol. 8, n. 2, pp. 164–73, 2000.

[63] J. P. Donoghue, «Connecting cortex to machines: Recent advances in
brain interfaces.,» *Nature Neuroscience,* vol. 5, pp. 1085–8, 2002.

[64] T. N. Lal, T. Hinterberger, G. Widman, M. Schröder, H. J., W.
Rosenstiel, C. E. Elger, B. Schölkopf e B. Birbaumer, «Methods towards invasive human brain computer interfaces,» *Advances
in Neural Information Processing Systems,* vol. 17, pp. 737–44,
2005.

[65] A. B. Schwartz, D. M. Taylor e S. I. H. Tillery, «Extraction algorithms for cortical control of arm prosthetics,» *Current Opinion in
Neurobiology,* vol. 11, n. 6, pp. 701–8, 2009.

[66] S. M. Coyle, T. E. Ward e C. M. Markham, «Brain-computer interface
using a simplified functional near-infrared spectroscopy system,»
Journal of Neural Engineering, vol. 4, n. 3, pp. 219–26, 2007.

[67] L. Kauhanen, T. Nykopp, J. Lehtonen, P. Jylänki, J. Heikkonen, P.
Rantanen, H. Alaranta e M. Sams, «EEG and MEG brain-computer
interface for tetraplegic patients,» *IEEE Transactions on Neural
Systems and Rehabilitation Engineering,* vol. 14, n. 2, pp. 190–3,
2006.

[68] J. Mellinger, G. Schalk, C. Braun, H. Preissl, W. Rosenstiel, N.
Birbaumer e A. Kübler, «An MEG-based brain-computer interface
(BCI),» *Neuroimage,* vol. 36, n. 3, pp. 581–93, 2007.

[69] R. Sitaram, A. Caria, R. Veit, T. Gaber, G. Rota, A. Kübler e N.
Birbaumer, «fMRI brain-computer interface: A tool for neuroscientific

research and treatment,» *Computational Intelligence and Neuroscience,* vol. 2007, p. 25487, 2007.

[70] N. Weiskopf, K. Mathiak, S. W. Bock, F. Scharnowski, R. Veit, W. Grodd, R. Goebel e N. Birbaumer, «Principles of a brain-computer interface (BCI) based on real-time functional magnetic resonance imaging (fMRI),» *IEEE Transactions on Biomedical Engineering,* vol. 51, n. 6, pp. 966–70, 2004.

[71] S. Yoo, T. Fairneny, N. Chen, S. Choo, L. P. Panych, H. Park, S. Lee e F. A. Jolesz, «Brain–computer interface using fMRI: Spatial navigation by thoughts,» *Neuroreport,* vol. 15, n. 10, pp. 1591–5, 2004.

[72] G. Vanacker, J. d. R. Millán, E. Lew, P. W. Ferrez, F. Galán, J. Philips, H. van Brussel e M. Nuttin, «Context-based filtering for assisted brain-actuated wheelchair driving,» *Computational Intelligence and Neuroscience,* vol. 2007, pp. 1–12, 2007.

[73] J. d. R. Millán, P. W. Ferrez, F. Galán, E. Lew e R. Chavarriaga, «Non-invasive brain-machine interaction,» *International Journal of Pattern Recognition and Artificial Intelligence,* vol. 2, n. 5, pp. 959–72, 2008.

[74] T. Carlson, L. Tonin, S. Perdikis, R. Leeb e J. d. R. Millán, «A hybrid BCI for enhanced control of a telepresence robot,» in *35th Annual International Conference of the IEEE Engineering in Medicine and Biology Society,* 2013.

[75] L. Tonin, R. Leeb, M. Tavella, S. Perdikis e J. d. R. Millán, «A BCI-driven telepresence robot,» *International Journal of Bioelectromagnetism,* vol. 13, n. 3, pp. 125–6, 2011.

[76] P. R. Kennedy, M. T. Kirby, M. M. Moore, B. King e A. Mallory, «Computer control using human intracortical local field potentials,» *IEEE Transactions on Neural Systems and Rehabilitation Engineering,* vol. 12, n. 3, pp. 339–44, 2004.

[77] D. J. McFarland e J. R. Wolpaw, «Brain-Computer interface operation of robotic and prosthetic devices,» *Computer,* vol. 41, n. 10, pp. 52–6, 2008.

[78] J. Lv, Y. Li e Z. Gu, «Decoding hand movement velocities from EEG signals during a continuous drawing task,» *Biomedical Engineering Online,* vol. 64, n. 9, 2010.

[79] A. Pressacco, R. Goodman, L. W. Forrester e J. L. Contreras-Vidal, «Neural decoding of treadmill walking from non-invasive, electroencephalographic (EEG) signals,» *Journal of Neurophysiology,* vol. 106, n. 4, pp. 1875–87, 2011.

[80] T. J. Bradberry, R. J. Gentili e J. L. Contreras-Vidal, «Reconstructing three-dimensional hand movements from noninvasive electroencephalographic signals,» *Journal of Neuroscience,* vol. 30, pp. 3432–7, 2010.

[81] H. A. Agashe e J. L. Contreras-Vidal, «Reconstructing hand kinematics during reach to grasp movements from electroencephalographic signals,» in *33rd conference of the IEEE Engineering in Medicine and Biology Society,* 2011.

[82] R. Poli e M. Salvaris, «Comment on fast attainment of computer cursor control with noninvasively acquired brain signals,» *Journal of Neural Engineering,* vol. 8, p. 058001, 2011.

[83] T. J. Bradberry, R. J. Gentili e J. L. Contreras-Vidal, «Reply to comment on 'fast attainment of computer cursor control with noninvasively acquired brain signals',» *Journal of Neural Engineering,* vol. 8, p. 058002, 2011.

[84] J. R. Wolpaw e D. J. V. T. M. McFarland, «Brain-computer interfaces for communication and control,» *Clinical of Neurophysiology,* n. 113, pp. 767–91, 2002.

[85] N. Birbaumer, «Slow cortical potentials: Plasticity, operant control, and behavioral effects,» *The Neuroscientist,* vol. 5, n. 2, pp. 74–8, 1999.

[86] N. Birbaumer, «Breaking the silence: Brain-computer interfaces (BCI) for communication and motor control,» *Psychophysiology,* vol. 43, n. 6, pp. 517–32, 2006.

[87] N. Birbaumer, N. Ghanayim, T. Hinterberger, I. Iversen, B. Kotchoubey, A. Kübler, J. Perelmouter, E. Taub e H. Flor, «A spelling device for the paralysed,» *Nature,* vol. 398, n. 6725, pp. 297–8, 1999.

[88] J. Perelmouter e N. Birbaumer, «A binary spelling interface with random errors,» *IEEE Transactions on Rehabilitation Engineering,* vol. 8, n. 2, pp. 227–32, 2000.

[89] G. Garipelli, R. Chavarriaga e J. d. R. Millán, «Single trial analysis of slow cortical potentials: A study on anticipation related potentials,» *Journal of Neural Engineering,* vol. 10, n. 3, p. 036014, 2013.

[90] G. Pfurtscheller e F. H. Lopes da Silva, «Event-related EEG/MEG synchronization and desynchronization: Basic principles,» *Clinical Neurophysiology,* vol. 110, n. 11, p. 1842–57, 1999.

[91] G. Pfurtscheller, C. Brunner, A. Schlögl e F. H. Lopes da Silva, «μ rhythm (de)synchronization and EEG single-trial classification of

different motor imagery tasks,» *Neuroimage,* vol. 31, n. 1, pp. 153–9, 2006.

[92] R. Leeb, S. Perdikis, L. Tonin, A. Biasiucci, M. Tavella, M. Creatura, A. Molina, A. Al-Khodairy, T. Carlson e J. d. R. Millán, «Transferring brain-computer interfaces beyond the laboratory: successful application control for motor-disabled users,» *Artificial Intelligence in Medicine,* vol. 59, n. 2, pp. 121–32, 2013.

[93] G. Pfurtscheller, C. Guger, G. R. Müller-Putz, G. Krausz e C. Neuper, «Brain oscillations control hand orthosis in a tetraplegic,» *Neuroscience Letters,* vol. 292, n. 3, pp. 211–4, 2000.

[94] G. R. Müller-Putz, C. Breitwieser, F. Cincotti, R. Leeb, M. Schreuder, F. Leotta, M. Tavella, L. Bianchi, A. Kreilinger, A. Ramsay, M. Rohm, M. Sagebaum, L. Tonin, C. Neuper e J. d. R. Millán, «Tools for Brain-computer interaction: A general concept for a hybrid BCI,» *Frontiers in Neuroinformatics,* vol. 5, n. 11, p. 30, 2011.

[95] B. Z. Allison, C. Brunner, V. Kaiser, G. R. Müller-Putz, C. Neupere G. Pfurtscheller, «Toward a hybrid brain-computer interface based on imagined movement and visual attention,» *Journal of Neural Engineering,* vol. 7, n. 2, p. 26007, 2010.

[96] P. Brunner, S. Joshi, S. Briskin, J. R. Wolpaw, H. Bischof e G. Schalk, «Does the P300 speller depend on eye gaze?,» *Journal of Neural Engineering,* vol. 7, n. 5, p. 056013, 2010.

[97] G. Pfurtscheller, B. Z. Allison, C. Brunner, G. Bauernfeind, T. Solis-Escalante, R. Scherer, T. O. Zander, G. Muller-Putz, C. Neuper e N. Birbaumer, «The hybrid BCI,» *Frontiers in Neuroscience,* vol. 4, n. 3, p. 42, 2010.

[98] P. Horki, T. Solis-Escalante, C. Neuper e G. Müller-Putz, «Combined motor imagery and SSVEP based BCI control of a 2 DoF artificial upper limb,» *Medical and Biological Engineering and Computing,* vol. 5, n. 49, pp. 567–77, 2011.

[99] F. Flemisch, C. Adams, S. R. Conway, K. H. Goodrich, M. T. Palmer e P. C. Schutte, «The H-metaphor as a guideline for vehicle automation and interaction,» *Hampton,* 2003.

[100] T. Carlson e J. d. R. Millán, «Brain-Controlled Wheelchairs: A Robotic Architecture,» *IEEE Robotics and Automation Magazine,* vol. 13, n. 1, pp. 65–73, 2013.

[101] T. Carlson, L. Tonin, S. Perdikis, R. Leeb e J. d. R. Millán, «A hybrid BCI for enhanced control of a telepresence robot,» in *35th Annual*

International Conference of the IEEE Engineering in Medicine and Biology Society, 2013.

[102] K. Goodrich, P. Schutte, F. Flemisch e R. Williams, «Application of the H-Mode, A Design and Interaction Concept for Highly Automated Vehicles, to Aircraft,» in *25th Digital Avionics Systems Conference,* 2006.

Biographies

R. Bortoletto received his Master degree from the University of Padova in 2011 under the guidance of Prof. Enrico Pagello. In 2012, he worked at IT+Robotics s.r.l. as software developer. Since January 2013, he is a PhD Student at the Intelligent Autonomous Systems Lab (IAS-Lab) of the Department of Information Engineering of the University of Padova. His research is currently focused in exploring new EMG-informed Neuromusculoskeletal Models for biomechanical analysis of human movement in neuromotor rehabilitation.

L. Tonin received his Ph.D. at the École Polytechnique Fédérale de Lausanne (EPFL, Lausanne, Switzerland) in 2013. Since October 2013 is working as Post-Doc at the Intelligent Autonomous System laboratory (IAS-Lab) in the University of Padua. In March 2014 he was co-founder and board member of EXiMotion s.r.l., a start-up company focused on research and development of Assistive Technology. His research is currently focused in exploring a novel Brain-Computer Interface based on Covert Visuospatial Attention for control and rehabilitation purposes. In particular, he is investigating the role of BCI in the framework of cognitive rehabilitation from spatial neglect syndrome in post-stroke patients. He published more than 30 articles in international journals and conferences and he participated as invited speaker to several international events and seminars.

Prof. E. Pagello received a "Laurea in Ingegneria Elettronica" at the University of Padova in 1972. He is a Full Professor of Computer Science since 2002, at the Department of Information Engineering, where he has been an Associate Professor since 1983. From '72 till '83, he has been a Research Associate of the National Research Council of Italy, where now he is a part-time collaborator. During 77–78, he was a Visiting Scholar at the Lab. of A. I. of Stanford University. He is a Fellow of the University of Tokyo, where has visited regularly the Department of Precision Engineering, since 1994. He has been a President of the Intelligent Autonomous Society, and a Vice-President of the RoboCup Int. Federation. His research interests are on applying Artificial Intelligence to Robotics.

Prof. E. Menegatti is Associate Professor of Computer Science in the School of Engineering of University of Padova. He graduated in Physics (1998), he has a Master of Science in Artificial Intelligence at the University of Edinburgh (UK) (2000) and a Ph.D. in Computer Science (2003). His research interests are in the field of Robot Vision Perception. In 2002, he was Visiting Researcher at University of Wakayama (JP), in 2004 he was at Georgia Institute of Technology (US) and at University of Osaka (JN). In 2005, Menegatti started IT+Robotics s.r.l., a Spin-off company of the Univ. of Padova, active in the field of intelligent industrial robotics. Menegatti has served as a Program Committee member for several conferences and chaired several workshops. In 2014, he is General Chair of the 13th Intl. Conf. on Intelligent Autonomous Systems (IAS-13). Menegatti was invited editor for three journal special issues, got two best paper awards and nominations, and published more than 25 journal papers and more than 100 conference papers.

4

Gaze Tracking, Facial Orientation Determination, Face and Emotion Recognition in 3D Space for Neurorehabilitation Applications

Krasimir Tonchev, Stanislav Panev, Agata Manolova, Nikolay Neshov,
Ognian Boumbarov and Vladimir Poulkov

Faculty of Telecommunications, Technical University of Sofia, Bulgaria
Corresponding authors: {k_tonchev, s_panev, amanolova, nneshov, olb, vkp}
@tu-sofia.bg

Abstract

Adaptive and interactive mental engagement combined with positive emotional state is requirement for an optimal outcome of the neurorehabilitation process for patients with brain damage usually caused by traumatic brain injury (TBI), stroke or brain diseases such as cancer, epilepsy and Alzheimer's disease. We propose a framework using an active multicamera system in 3D space for human gaze tracking and face orientation determination combined with 3D face and emotion recognition based on multiple kernel learning. This type of framework is suitable for being part of assistive medical system for neurorehabilitation of patients with TBI. The experiments show comparable recognition rate (79.91%) to similar algorithms recognizing the emotional states of the subjects.

Keywords: active multi-camera system, gaze tracking, face orientation, facial expression recognition, multiple kernel learning.

4.1 Introduction

Brain damage in humans can have multiple causes such as brain injuries as a result of motor vehicle crashes, sports injuries or simple falls on the

Neuro-Rehabilitation with Brain Interface, 51–88.

playground, at work or in the home or diseases such stroke, cancer, epilepsy, Alzheimer's disease, Parkinson's disease, multiple sclerosis and others. According to [1], there are two million TBI cases each year (one every 15 seconds), and in every 5 minutes, someone becomes permanently disabled due a head injury. A total of 70,000 to 90,000 of those who survive will have lifelong disabilities and will require 5 to 10 years of intensive services such as vocational rehabilitation and physical therapy. TBI is an enormous burden for the public health services and has an important socio-economic weight throughout the world. Although correct statistical data are difficult to gather, it is estimated that in the USA, around 5.3 million people are living with a TBI-related disability [2], and in the European Union, approximately 7.7 million people who have experienced a TBI have disabilities [3]. Such types of injury to the brain commonly lead to neurocognitive deficits (e.g. impaired attention, inability to form visuospatial associations or poor executive function) and psychological health issues; for example, 30–70% of TBI survivors develop depression.

The brain areas that are typically injured (caused by TBI, stroke and tumours) are the frontal lobes. This fact is due to the frontal lobe being positioned in the most forward portion of the brain, just underneath the forehead. This part of the brain occupies the largest portion of the human brain, approximately one-third of the entire brain. The frontal lobe manages the control of thoughts, emotions and actions. Its three subregions are responsible for different things such as the prefrontal cortex is responsible for complex thought and emotional expression; the inferior frontal cortex controls smell, instincts and raw emotions; and the posterior (back) frontal lobe controls motor functions, including speech. The observed problems following frontal lobe damage may include loss of simple movement of various body parts (paralysis), inability to plan a sequence of complex movements needed to complete multistepped tasks, loss of flexibility in thinking, inability to focus on task, impulsivity, poor planning, inadequate responses [5], mood changes, difficulty controlling behaviour – poor anger management [4], and difficulty with problem-solving [6].

The depression and anxiety combined with all the above-mentioned problems can affect the person's interpersonal relationships and contribute to poor communicability, social and vocational integration and may lead to alienation, long-term placement in an institutional setting and even to suicide. Yearly cost from TBI in terms of hospitalization and rehabilitation costs in Europe alone exceeds 100 billion euros (this sum does not include indirect costs to society as well as to families, but does include costs associated with lost earnings and productivity, as well as the costs linked to providing social services). The statistics show a steep increase in the incidence of TBIs, with an increase of 21% over the last five years [7]. According to the TBIcare project [7], despite this TBI, rehabilitation has been seriously underrepresented in medical and engineering R&D efforts compared to many other, less significant health problems. Standard rehabilitation methods currently used are struggling to provide consistently meaningful improvements to patient abilities. To improve the chances of the patient's recovery, new methodologies and technologies from different fields of medicine, psychology and engineering must be incorporated in the rehabilitation process.

Neurorehabilitation is a relatively new field which involves multiple disciplines that combine series of therapies from the psychological to occupational, teaching or retraining patients on mobility skills, communication processes and other aspects of that person's daily routine [8]. This complex medical process is aimed at reducing the nervous system-related impairments, disabilities and handicaps, and ultimately enhancing the quality of life for persons after brain damage. The use of emerging technologies such as robotics, virtual reality and brain–computer interfaces for enhancing their user's independence will lead to better living by allowing remote assistance from the medical personnel, which in turn may reduce the stress of a visit to the hospital [9] or the pain in patients with mobility impairments [10]. The patients will benefit from the possibility of remote interaction with their doctors without going outside their comfort zone and also carry out the training from their home, under remote supervision, reducing the cost to the healthcare system. For doctors, these types of assistive rehabilitation systems provide online remote monitoring of both the rehabilitation process and the physiological state of the patient [8]. So it is very important for the medical personnel to be able to follow the emotional state of the patient because it plays a big role in the rehabilitation process.

In the therapeutic context, common behavioural symptoms may include resistance, verbal or physical aggressiveness, agitation and refusal to engage in the therapeutic session. To help patients achieve rehabilitative goals, it is important to understand why such behaviours may occur and ways to minimize or manage them [11]. So an automatic facial emotion recognition framework is really important part of human–computer interaction for assistive medical systems.

According to the face recognition vendor test (FRVT) report on the performance of face identification algorithms from 2014 [54], the face recognition accuracy is very strongly dependent on the proposed algorithm and, more specifically, on the developer of the algorithm. Recognition error rates in a particular scenario (different types of images taken in different conditions) range from a few percent up to beyond fifty percent [54]. The development of new technologies, especially those involving 3D facial scans, has led to overall increase in accuracy for all the developers who have submitted algorithms for the FRVT during 2010 and 2013. But still the accuracy of facial recognition implementations varies greatly across the industry and the majority of the participating commercial vendors have not disclosed the details of their methods, so it is difficult to assess what is distinctive about their algorithms. According to the FRVT's comparative tests of the face recognition capabilities of humans and software, six out of seven automatic face recognition algorithms were comparable to or better than human recognition [55].

At this stage of research, the degree of accuracy in facial emotion recognition has not been brought to a level high enough to permit its widespread efficient use across the many practical applications that require such action. Some of the challenges to overcome are connected to the fact that posed expressions, as used in most studies, are not natural and therefore cannot represent 100% the real emotional state. Also, we need to consider the lack of rotational movement freedom. Most of the proposed algorithms work very well with frontal images, but upon rotating the head more than 20 degrees, there have been problems due to lost facial information [55].

The goal of the current study was to propose a framework using an active multicamera system in 3D space for human gaze tracking and face orientation determination combined with 3D face and emotion recognition based on multiple kernel learning. This framework has the potential to be implemented in an assistive medical system for neurorehabilitation with the goal to help the medical personnel from the remote location to assess and manage the behaviour by modifying contributing factors that place too much demand or press on the patient during the training process. Such factors may include the physical environment (e.g. auditory and visual distractions), the social environment (e.g. communication style of informal/formal caregivers) or factors that are modifiable but which are internal to the individual themselves (e.g. discomfort, pain, fatigue).

The rest of the chapter is organized as follows: in the next section, we present a gaze tracking and face orientation determination with an active multicamera system with Kinect sensor. In Section 4.3, we present the approach for 3D face recognition with multiple kernel learning. In that section, we describe the different preprocessing steps of the 3D image data and classification process. In Section 4.4, we present a 3D expression recognition algorithm with the experimental results for the classification of the processed data and comparison with similar works. Finally, in Sections 4.5 and 4.6, we will present our future plans to develop these methods in the framework of assistive medical system for neurorehabilitation.

4.2 Gaze Tracking and Face Orientation Determination with an Active Multicamera System with Kinect Sensor

Using the information from the gaze as a form of input can enable a computer system to gain more contextual information about the user's task at hand, which in turn can help to design interfaces which are more interactive and intelligent. The existing gaze tracking techniques are broadly classified into intrusive and non-intrusive.

The intrusive techniques require attachments around the eye to determine the gaze. These include search coils, electrooculography [14], contact lens and head-mounted devices. Non-intrusive techniques use video cameras under infrared or natural light sources. The non-intrusive or video-based techniques are classified into two categories, appearance based and model based. Appearance-based approaches directly treat an eye image as a high-dimensional feature. In [17], the authors use the image contents as to map directly to the screen coordinates. These methods require several significant calibration points to infer the gaze direction from the images. The analysis of the images at calibration points is important for gaze estimation. Baluja and Pomerleau use a neural network to learn a mapping function between eye images and gaze points (display coordinates) using 2,000 training samples [13]. Tan et al. take a local interpolation approach to estimate unknown gaze points from 252 relatively sparse samples [27]. Recently, Williams et al. proposed a novel regression method called S^3GP (sparse, semi-supervised Gaussian process) and applied it to the gaze estimation task with partially labelled (16 of 80) training samples [28]. Appearance-based approaches can make the system less restrictive and can also be very robust even when used with relatively low-resolution cameras. The appearance models are used for tracking smaller eye movements compared to the size of the object. Model-based approaches use an explicit geometric model of the eye and estimate its gaze direction using geometric eye features. For example, one typical feature is the pupil–glint vector ([18, 20]), the relative position of the pupil centre and the specular reflection of a light source, pupil–corneal reflection. Model-based approaches typically need to precisely locate small features on the eye using a high-resolution image and often require additional light sources but can be very accurate. The local gaze features include pupil and limbus position, iris centre, eye corner and inner eye boundary and sclera region. The global gaze features are face skin colour, interpupil distance, ratio of average intensity, shapes, sizes of both the pupils and orientation of pupil ellipse with respect to face pose [21].

In this section, we present an improved framework for determining the direction of human gaze with an active multicamera system. We employ a fixed camera in order to determine the position of the human face and its features, most importantly the eyes. By means of the supervised descent method (SDM) for minimizing a nonlinear least squares (NLS) function, we can compute correctly the position of the two eyes using 6 landmarks for each eye and the position of the head. Then, an active pan–tilt camera is oriented to one of the users' eyes, and in this way, a high-precision gaze direction determination is accomplished.

4.2.1 System Overview

The proposed gaze tracking system consists of the components shown in Figure 4.1. The purpose of the system described in this chapter is to estimate the gaze direction of a user which is standing in front of a computer screen. To accomplish that task with a video camera, as precise as possible, a larger-scale image of the user's eye(s) is needed. This can be achieved by mounting a telephoto lens on the camera, but such an action will narrow the camera's field of view and respectively will constrain the user's movements in order to keep the image of its eye(s) inside the camera frame. To overcome this problem, a multicamera gaze tracking system is constructed, which consists mainly of two parts – a fixed 3D wide-angle depth sensor and a movable telephoto camera which is mounted on an active pan–tilt unit (PTU) [58]. The wide-angle 3D sensor is used to detect the 3D position of the user's face and eyes in a global coordinate frame. This information is needed by the PTU control system to estimate the angles of rotation of the telephoto camera, so it will be pointed towards the user's eye(s). Then, the gaze direction can be estimated by extracting eye's features from the large-scale image and applying them to a human eye geometrical model.

The 3D depth sensor Kinect is equipped with two imaging devices – a colour RGB camera and a greyscale camera which works in the IR spectrum. In combination with an IR laser projector, the 3D range sensor is constructed. The disparity map and the pixel registration between it and the image from the RGB camera are automatically calculated.

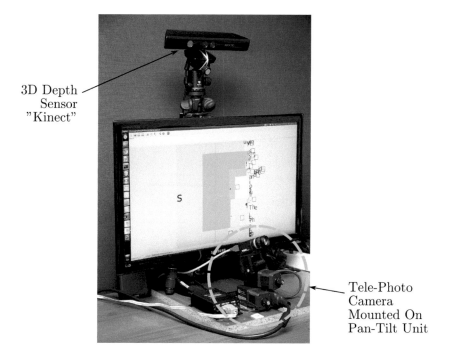

Figure 4.1 System overview

As an output from the Kinect, we get a depth map, and by employing its intrinsic camera parameters, such as lens focal length and principal point coordinates, the so-called 3D point cloud is estimated. This information will be used later on for 3D face and eye tracking as mentioned earlier.

The block diagram shown in Figure 4.2 describes the main stages and their consistency for processing the information of the two data sources (marked in blue) until the final aim is achieved. All of them are described in the rest of this section.

4.2.2 System Geometrical Model and Calibration

The interpretation of the data retrieved from the Kinect [58], in order to control the active camera, requires the employment of a 3D system geometrical model which will be the backbone of all the calculations concerning the user's face and eye tracking.

The common system geometrical model is depicted in Figure 4.3. The depth sensor and the video camera have their own coordinate frames denoted as $\{K\} \equiv \{O_K, X_K, Y_K, Z_K\}$ and $\{C\} \equiv \{O_C, X_C, Y_C, Z_C\}$, respectively,

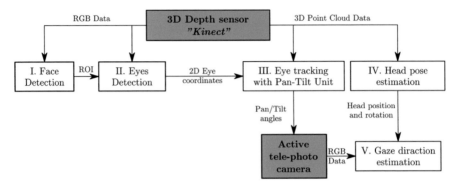

Figure 4.2   Block diagram of the system

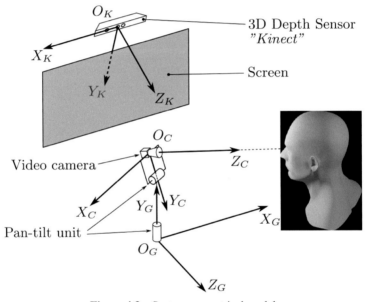

Figure 4.3   System geometrical model

and $\{G\} \equiv \{O_G, X_G, Y_G, Z_G\}$ is the fixed global coordinate frame which is used as a reference for all other frames. The transformation of the coordinates from one frame to another is realized by 4×4 homogeneous transformation matrices $T \in SE(4.3)$ [14]:

$$T = \begin{bmatrix} R & t \\ 0 & 1 \end{bmatrix}, \tag{4.1}$$

where R is the rotation matrix and t is the translation vector.

As can be learned from Figure 4.1 and Figure 4.3, there is no obvious relation between the global and the Kinect's coordinate frames ($\{G\}$ and $\{K\}$) which can be expressed analytically as a transformation matrix. That is why a template-based calibration procedure is performed to determine the transformation of the coordinates between these two coordinate frames and it is described below in Section 4.2.2.2.

In contrast to the $\{G\}$ and $\{K\}$ relation, the relation between the frames of the telephoto camera $\{C\}$ and global frame $\{G\}$ is clear; it depends only on the pan and tilt angles (θ and φ, respectively (Figure 4.4)) and some physical dimensions. The derivation of this transformation is detailed in the following section.

4.2.2.1 Modelling the pan–tilt unit

For the goal of our research, a pan–tilt unit model is developed, based on the existing PTU-D46-17 system produced by the company Directed Perception (now known as FLIR Motion Control Systems, Inc.). This unit is distinguished by its precision ($\approx 0.001°$), build quality and ease of use through its controller.

The derivation of the matrix for transforming the coordinates between the telephoto camera and the global frames is based on the kinematic chain of

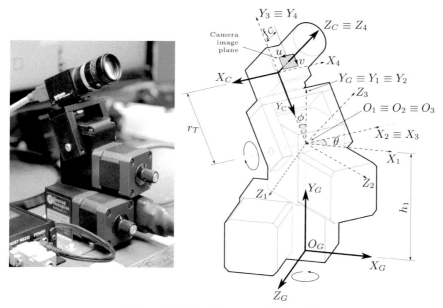

Figure 4.4   Pan–tilt geometrical model

the PTU, which is shown in Figure 4.4. The relation between $\{C\}$ and $\{G\}$ can be expressed as five simple intermediate transformations – translations or rotations about a single axis. Let all these transformations be depicted by a numeric index which will distinguish them from the main frames. The first coordinate transformation is a translation along the Y_G axis at a height equal to h_1, which is a physical parameter of the PTU. This kind of transformation is expressed by the following matrix:

$$T_{G\to 1} = \begin{bmatrix} 1 & 0 & 0 & 0 \\ 0 & 1 & 0 & -h_1 \\ 0 & 0 & 1 & 0 \\ 0 & 0 & 0 & 1 \end{bmatrix}. \tag{4.2}$$

The second transformation is simple rotation around the axis $Y_1 \equiv Y_G$ by angle θ which represents the *pan* motion:

$$T_{1\to 2} = \begin{bmatrix} \cos(\theta) & 0 & \sin(\theta) & 0 \\ 0 & 1 & 0 & 0 \\ -\sin(\theta) & 0 & \cos(\theta) & 0 \\ 0 & 0 & 0 & 1 \end{bmatrix}. \tag{4.3}$$

The third transformation is simple rotation around $X_2 \equiv X_3$ by angle φ which represents the *tilt* motion of the PTU:

$$T_{2\to 3} = \begin{bmatrix} 1 & 0 & 0 & 0 \\ 0 & \cos(\varphi) & -\sin(\varphi) & 0 \\ 0 & \sin(\varphi) & \cos(\varphi) & 0 \\ 0 & 0 & 0 & 1 \end{bmatrix}. \tag{4.4}$$

The transformation from frame $\{3\}$ and frame $\{4\}$ is a translation along axis $Y_3 \equiv Y_4$ by the PTU eccentricity radius r_T:

$$T_{3\to 4} = \begin{bmatrix} 1 & 0 & 0 & 0 \\ 0 & 1 & 0 & -r_T \\ 0 & 0 & 1 & 0 \\ 0 & 0 & 0 & 1 \end{bmatrix}. \tag{4.5}$$

The final simple transformation consists of a rotation around the axis $Z_4 \equiv Z_G$ by π radians. This rotation is done to align the camera coordinate frame to the image coordinate frame – $(u,\ v)$:

$$T_{4 \to C} = \begin{bmatrix} \cos(\pi) & -\sin(\pi) & 0 & 0 \\ \sin(\pi) & \cos(\pi) & 0 & 0 \\ 0 & 0 & 1 & 0 \\ 0 & 0 & 0 & 1 \end{bmatrix}. \tag{4.6}$$

By multiplying (4.2), (4.3), (4.4), (4.5) and (4.6), the geometrical model of the PTU is constructed and has the following form:

$$T_{G \to C} = T_{4 \to C}.T_{3 \to 4}.T_{2 \to 3}.T_{1 \to 2}.T_{G \to 1} =$$

$$= \begin{bmatrix} -c_\theta & 0 & -s_\theta & 0 \\ -s_\theta s_\varphi & -c_\varphi & c_\theta s_\varphi & r_T + h_1 c_\varphi \\ -s_\theta c_\varphi & s_\varphi & c_\theta c_\varphi & -h_1 s_\varphi \\ 0 & 0 & 0 & 1 \end{bmatrix}, \tag{4.7}$$

where with s and c are depicted the *sin* and *cos* functions, respectively. The reverse transformation $\{C\} \to \{G\}$ is achieved with the inverse of the above matrix:

$$T_{C \to G} = T_{G \to C}^{-1}. \tag{4.8}$$

4.2.2.2 Estimating the transformation $\{G\} \leftrightarrow \{K\}$

As mentioned above, there is no clear kinematic relation between the depth sensors and the global coordinate frames. That is why a calibration technique has been developed to estimate the transformation matrix between these two frames, which is based on the assumption that they are fixed in space and the transformation between them is constant:

$$T_{G \to K} = T_{K \to G}^{-1} = const. \tag{4.9}$$

As a calibration target, a simple pattern is used as shown in Figure 4.5a. The algorithm for detecting such kind of targets is based on the marker detection technique from [12], but it has been modified and the target corner detection accuracy is improved to subpixel level. As can be seen from the figure, the calibration target consists of a black square and inscribed in it, there is another white one. The algorithm [12] detects the contours of the squares and tracks them. As the calibration target is planar, its plane defines the *XY* plane of the 3D coordinate frame attached to it. The origin of this frame is placed in one of the corners of the inner (white) square. The little black marker marks the corner chosen for the origin of the coordinate system (Figure 4.5a).

In order to estimate the transformation matrix $T_{K \to G}$, the calibration target will be used as a reference coordinate frame which will give the relation

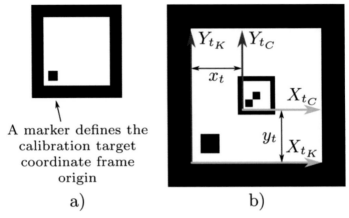

A marker defines the
calibration target
coordinate frame
origin

a)   b)

Figure 4.5   Calibration target

between the two imaging devices. This means that the target should be present
in the images of the two input devices during the calibration procedure. But
this is a problem because their fields of view (FOVs) are completely different.
That is why a combination of two calibration patterns with different sizes
corresponding to the FOVs of the cameras is used. Their coordinate frames
share a common *XY* plane (Figure 4.5b), and thus, the transformation between
them is described as translation only:

$$
T_{t_K \to t_C} = \begin{bmatrix} 1 & 0 & 0 & -x_t \\ 0 & 1 & 0 & -y_t \\ 0 & 0 & 1 & 0 \\ 0 & 0 & 0 & 1 \end{bmatrix},
\tag{4.10}
$$

where $\{t_K\}$ and $\{t_C\}$ are the coordinate frames of the two calibration targets
for the Kinect and for telephoto video camera, and x_t and y_t are the translations
between the two frames along the *X*- and the *Y*-axes (Figure 4.5b). As the
dimensions of the calibration target are known, the pose of the cameras can be
estimated with respect to the calibration target by employing an algorithm
for solving the perspective-n-point (PnP) problem. This will produce the
transformation matrices $T_{l_K \to K}$ and $T_{t_C \to C}$. Such kind of algorithms like
the iterative method based on Levenberg–Marquardt optimization or the P3P
solution in [15] are implemented as C/C++ libraries and can be used for
real-time applications.

Figure 4.6 shows the scheme for estimating the transformation between
the $\{K\}$ and $\{G\}$.

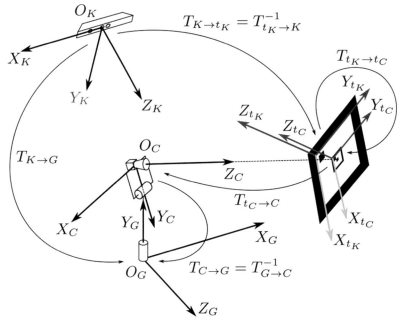

Figure 4.6 Estimation of $T_{G \to K}$ transformation matrix

From the figure, it is obvious that $T_{K \to G}$ can be expressed as four consequential transformations of the coordinates:

1. between the Kinect's frame to the frame of its calibration pattern ($\{K\} \to \{t_K\}$),
2. between the two calibration patterns $\{t_K\} \to \{t_C\}$,
3. between the telephoto camera calibration and its pattern $\{t_C\} \to \{C\}$
4. between the active camera and the global coordinate frame $\{C\} \to \{G\}$.

This is described by the following product:

$$T_{K \to G} = T_{C \to G}.T_{t_C \to C}.T_{t_K \to t_C}.T_{K \to t_K} = T_{G \to C}^{-1}.T_{t_c \to c}.T_{t_K \to t_C}.T_{t_K \to K}^{-1} \tag{4.11}$$

4.2.3 Facial Extraction and Tracking

The gaze direction estimations rely on detection of the eye corners' positions and the head pose (Section 4.2.4). The positions of the eye corners are estimated by employing a modern optimization technique, called supervised decent method (SDM), for aligning a face model, consisting of 48 landmarks,

Figure 4.7   The red outline depicts the result of the face detector [27]. Key points detected after applying the SDM method

to a face image [29]. The main advantage of SDM is that during the optimization process, none of the Jacobian and Hessian is calculated (in contrast to the Newton's optimization methods), which could be computationally expensive. This is achieved by learning a series of decent directions and rescaling factors such that a sequence of updates of the optimized function produced starting from the initial face model state (x_0) that converges to the manually aligned face model (x_*) in the training data. The x_0 are the landmarks' positions of the mean face given by a face detector algorithm [27]. After aligning the face model to the image of the face, the head pose is estimated. Figure 4.7 depicts the key points obtained after applying the SDM method.

Since each pixel from the RGB camera of the Kinect has a corresponding point from the point-cloud array of the same device, the 2D locations of the eyes, estimated as a mean of the eye landmarks' positions, are directly converted into 3D location in the $\{K\}$ coordinate frame. By using (4.11), they can be transformed into coordinates of the global frame $\{G\}$. Thus, the PTU active camera can be navigated to track the location of one of the eye with the help of (4.7).

4.2.4 Gaze Direction Estimation

A description of the geometrical eye model is given in order to estimate the gaze direction. We assume that the eyeball is spherical and the inner and outer eye corners have been estimated, in our case using SDM (Section 4.2.3). The algorithm is split into two steps [19]:

1. Estimate the centre and the radius of the eyeball in the image from the eye corners and the head pose.
2. Estimate the gaze direction from pupil location, the centre and the radius of the eyeball.

The first of these two steps requires the following anatomical constants (Figure 4.8):

- R_0: The radius of the eyeball in the image when the scale of the face is 1.
- (T_x, T_y): The offset in the image between the mid-point of the two eye corners and the centre of the eyeball.
- L: The depth of the centre of the eyeball relative to the plane containing the eye corners.

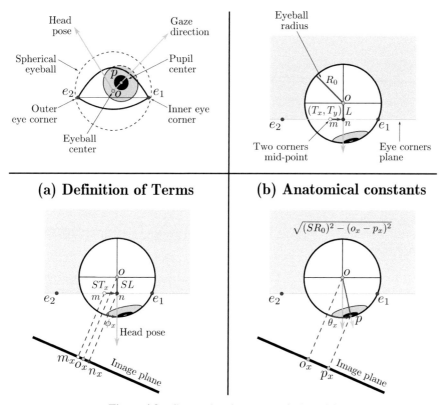

(a) Definition of Terms

(b) Anatomical constants

Figure 4.8 Gaze estimation geometrical model

Figure 4.9   Demonstration of aligning a model to an image of a face from the RGB camera of Kinect sensor by means of the SDM optimization algorithm. Consequentially by taking the information of the locations of the eyes in the image and applying it to the depth map of the Kinect, the 3D coordinates of them are estimated in the $\{K\}$ frame

4.2.5 Experimental Results

In Figure 4.9, an example is shown of aligning a face model to an actual face in the RGB image from Kinect sensor. By using the information about the eyes' locations in the colour image, their 3D position can be estimated by applying this information to the depth map of the sensor. Thus, the PTU is directed to one of the eyes and the gaze direction is estimated by the technique described in Section 4.2.4.

The facial orientation in the 3D space is done also with SDM and can be observed in Figure 4.9 at the upper left corner. The three axes show the position of the head in the space.

4.3 3D Face and Emotion Recognition with Multiple Kernel Learning

According to [30], it appears that 2D face recognition techniques have exhausted their potential as they stumble on inherent problems of their modality such as differences in pose, lighting, expressions and other characteristics that can vary between captures of a human face. In comparison with the 2D images, the 3D face scans have the advantages of being illumination invariant and capturing the true face surface. 3D face recognition is a challenging task with a large number of proposed solutions [31, 32]. With variations in pose and expression, the identification of a face scan based on 3D geometry is difficult.

A face-based biometric system consists of several subsystems: acquisition system performed by special devices (2D camera, 3D scanner or infrared camera), preprocessing unit, feature extraction unit, database and a recognition unit. In our scenario, an acquisition framework has been presented in Section 4.2.

In our research, we have at our disposal 3D image data sets that are acquired by the 3D scanner and not by the Kinect infrared camera. But for future development, 3D facial image data sets from the Kinect sensor will be gathered.

One common technique on 3D object recognition is based on the correspondence among scene points and model points in order to perform the recognition and to determine the object pose and location [33]. Among the 3D free-form object descriptors to represent objects is the curvature of the local surface evaluated in each point, which is characterized by the directions in which the normal of the surface changes more or less quickly [34]. In [34], a set of twelve 3D features extracted from segmented regions using curvature properties of the surface were experimented for face recognition using a database of 8 individuals and 3 images per individual obtaining 95,5% recognition rate providing a previous 100% correct feature extraction.

Another popular approach to representing the face is the range image representation where the 3D point cloud is represented as a 2D image. The popularity of this approach is due to the many readily available methods for 2D facial recognition [36]. Also, range images are robust to the change of colour and illumination, which causes a significant problem in face recognition using 2D intensity images.

In this section, we propose a framework for 3D face and emotion recognition based on a) preprocessing, b) a couple of geometric features with corresponding kernel functions and c) multiple kernel learning. The rest of the section is organized as follows: In Paragraph 4.3.1, we present the preprocessing of the data and the feature extraction step; in Paragraph 4.3.2, we present the SimpleMKL algorithm and the proposed framework; in Paragraph 4.3.3, we provide a detailed result of our experiments with connection to the 3D face recognition; and finally, in Paragraph 4.3.4, we present the results for the facial expression recognition.

4.3.1 Preprocessing and Feature Extraction

4.3.1.1 Point cloud filtering

Compared to widely available CMOS and CCD technologies for capturing 2D images, the technology for 3D scanning is still immature in many aspects. There are multiple approaches to acquiring 3D data, but all of them have drawbacks. The data acquired with 3D scanners contain undesired noise usually in the form of erroneous measurements or missing measurements. In the first case, this noise appears as "spikes" in the data and in the second appears as "holes". An example is illustrated in Figure 4.10.

Usually, samples are missing because the scanner fails to read the reflection of the laser beam at a certain point, while the spikelike measurements are usually caused by reflection of wet surfaces such as the eyes [35]. The spikes exist because the laser beam is reflected by a glossy object such as wet skin areas. The holes on the other hand can be seen when the laser beam of the scanner is not reflected. Such situations can rise, for example, when the mouth of the person is open or the pupil of the eye is wide open. It may heavily influence the recognition processes, so a preprocessing step is required.

To cope with these issues, we apply preprocessing in three steps: face extraction, median filtration and smoothing. Median filtration is used to remove the spikelike measurements. The smoothing is performed using cubic interpolation. All missing points are filled by interpolation based on the closest points. Since the interpolation is applied in a least squares manner, a Gaussian noise is reduced also.

The scanning process captures the face but also the region around the face including hair, neck and chests. These body parts do not contain information

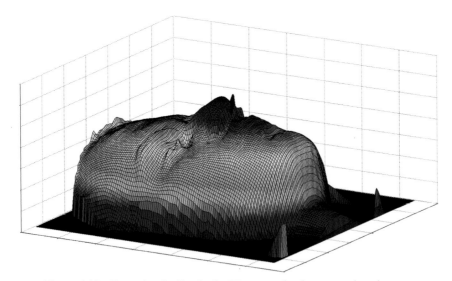

Figure 4.10   Example of spikes in the 3D scan under the nose and on the eye

relevant to the face recognition process and need to be removed. We follow the common approach for extracting the face region by selecting all points located inside a sphere with radius R_s centred at the noise tip (assuming the nose tip location is known) [4k].

To cope with these drawbacks, we apply 3D median filtering of size N_{med} elements. The steps to perform this filtration are as follows: first, for each point $x_i = [x_i, y_i, z_i]^T$ of the point cloud data, the nearest N_{med} points are selected by calculating the Euclidean distance using only the x and y coordinates. Next, the z of these N_{med} points are arranged in increasing order, and the value at index $\lfloor N_{med}/2 \rfloor + 1$ is selected. This value represents the median value of the selected N_{med} points, closest to x_i. As a final step, the z_i value of x_i is replaced with the median value.

This filtration removes the "spikes" and the "holes" and also the noise induced by the scanner.

The next step in the preprocessing stage is face extraction (Figure 4.11). This step is needed because the scanner captures data of body parts such as neck, shoulders and hair. This step can also be considered as face registration step because all faces are aligned to the same coordinate centre.

4.3.1.2 Surface normal feature extraction

Surface normals encode the rate of change of the face surface over local patches [36]. In this task, the surface normal is a three-component vector pointing outwards the facial surface at a particular point. Hence, for each point of the point cloud data, we can associate a feature represented by the surface normal at that point. We calculate the surface normals using principal component

Figure 4.11 Original 3D face data and preprocessed one

analysis (PCA). Given a zero mean data, represented in a matrix form $X = \{x_1, x_2, \ldots, x_N\}$, $x_i \in \mathbb{R}^n$, PCA calculates the eigenvalue decomposition of its sample covariance matrix $C = \frac{1}{N}XX^T$:

$$C = V^T \Lambda V, \tag{4.12}$$

where V are the eigenvectors and Λ are their corresponding eigenvalues. Each eigenvalue represents the variation of the data in the direction of its corresponding eigenvector. In order to extract the surface normals, as a first step, we apply PCA on local surface patches of size $N_{PCA} \times N_{PCA}$ centred on each data point. Then, the eigenvector with the smallest positive eigenvalue is extracted as a feature vector. Face surface and its corresponding surface normals are depicted in Figure 4.12.

4.3.1.3 Locally adaptive regression kernels (LARK) feature extraction

LARK is a technique for feature extraction recently proposed by [37]. It measures the distance between neighbouring data points (pixels) along the signal manifold, i.e. the geodesic distance. Geodesics represent the distance between two points on the surface while travelling without leaving the surface. More formally, this is the shortest curved line of the surface connecting

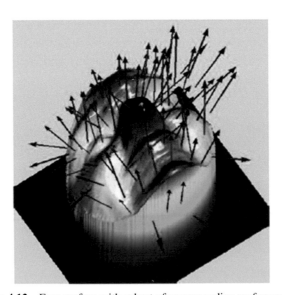

Figure 4.12   Face surface with subset of corresponding surface normals

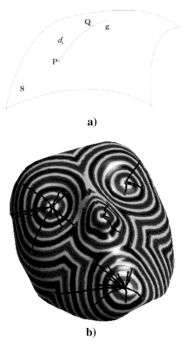

a)

b)

Figure 4.13   a) Definition of geodesic distance d_s of an arbitrary point Q. Geodesic path g is the minimum length curve connecting point Q and geodesic pole P. d_s is the length of g. b) Example for geodesic paths and circles defined over a 3D surface "nefertiti" scan with four starting points (geodesic poles), generated with the MATLAB code provided in [56]

two points [38] (see Figure 4.13a). This is achieved by exploiting the local self-similarity based on the gradients.

In [39], a method is proposed to extract features in a local region around certain point called locally adaptive regression kernels. Evaluating the geodesics can be done based on the elementary distance d_s. Since the point cloud represents a surface (the face surface) embedded in $\mathbb{R}^3$, it can be parametrized over the x and y coordinates $S(x, y) = \{x, y, z(x, y)\}$. Then, the local differential on a surface is calculated by [37]

$$d_s^2 = \Delta\mathbf{x}^T C \Delta\mathbf{x} + \Delta\mathbf{x}^T \mathbf{x}, \tag{4.13}$$

where $\Delta\mathbf{x} = [dx, dy]^T$, i.e. the differential in spatial coordinates. The matrix C is the structure tensor (the local gradient covariance matrix) given by

$$C = \begin{bmatrix} z_x^2 + 1 & z_x z_y \\ z_x z_y & z_y^2 + 1 \end{bmatrix}, \tag{4.14}$$

where z_x and z_y are the derivatives in x and y directions. The authors in [37] propose the following steps for calculating LARK:

1. For each data point, select a surrounding window W with size $N_{LARK} \times N_{LARK}$;
2. For each point in this window, calculate the local gradient covariance matrix (4.14);
3. Perform averaging on the calculated local covariance matrices over all points $p \in W$:

$$\overline{C} = \sum \left[\begin{array}{cc} z_x(p)^2 & z_x(p)z_y(p) \\ z_x(p)z_y(p) & z_y(p)^2 \end{array} \right]; \qquad (4.15)$$

4. Regularize the average local gradient covariance matrix $\overline{C}$ (for further details about the regularization, see [39]);
5. Calculate the self-similarity between the centre and its surrounding of the window W by

$$K\left(\overline{C}, \triangle \mathbf{x}_p\right) = \exp\left\{-\triangle\mathbf{x}_p^T \overline{C} \triangle\mathbf{x}_p\right\}. \qquad (4.16)$$

The Steps 4 and 5 are necessary because calculating the gradients (z_x, z_y) is very prone to noise. The last step of the above procedure defines LARK. Note that the term $\triangle\mathbf{x}^T\mathbf{x}$ defined in (4.13) is omitted since because it is data independent and its effect on a local window is trivial.

Extracting the LARK features is done by applying the above-defined procedure to each data point in the data set. For more convenient representation, each feature is arranged as a row vector and then all are stacked, ending up with a matrix of size $N \times N_{LARK}^2$, where N is the number of the available data points and N_{LARK}^2 is the number of elements in the window used for calculations. Usually, the LARK features are calculated for a subset of the data points.

In other words, LARK features represent an approximation of the geodesic distance between central point and its surrounding in a window, weighted by an exponential function. Hence, LARK features describe one of the intrinsic properties of the surface and carry information that can be used for classification.

4.3.1.4 Shape operator
The LARK features based on the geodesics showed promising results as indicated in [40]. Nevertheless, using the geodesics alone may be ambiguous. There are plenty of cases where different curves have the same geodesic

distance but are different in shape. In this case, additional features are necessary which, intuitively speaking, should provide information about the shape of the surface.

The most popular features, providing shape information in local region, used in the community working in 3D facial analysis are the mean **H** and Gaussian curvatures **K**. Moreover, these curvatures are the most important in general according to surface theory. There are multiple ways to introduce **H** and **K**, but as will become clear later on, we will use the one based on the shape operator. The Weingarten map, also called the shape operator, is a linear map from the tangent space of a surface at a point p to that same tangent space [41]. The operator is defined as the negative of the directional derivative of the unit surface normal n at p:

$$W_p(\boldsymbol{w}) = -\nabla_{\boldsymbol{w}} \boldsymbol{n}, \qquad (4.17)$$

where $\boldsymbol{w}$ is a tangent vector on the surface at $\boldsymbol{p}$. The Weingarten map is fundamental in analysing the surface's curvature. For a 3D surface, given an orthonormal basis for the tangent space at $\boldsymbol{p}$, W_p can be represented as a 2×2 symmetric matrix, whose components involve second (and possibly first) derivatives of that surface.

Deriving the operator in explicit form is beyond the scope of this work in which case we will use the final result [40]:

$$W_p = \frac{-1}{EG - F^2} \begin{bmatrix} e & f \\ f & g \end{bmatrix} \begin{bmatrix} E & -F \\ -F & G \end{bmatrix}, \qquad (4.18)$$

where E, F and G are the coefficients of the first fundamental form [40] and e, f and g are the coefficients of the second fundamental form, respectively. Notice that the denominator is the determinant of the second matrix. This expression allows for smoothing and regularizing the shape operator, thus avoiding numerical instabilities and reducing the noise influence. One important property of the operator, useful in computations, is its symmetry. Although the second fundamental form is not included explicitly, it can be assumed that its coefficients carry information about the extrinsic properties of the surface.

Calculating the mean and Gaussian curvatures using the eigenvalues of (4.18), k_1 and k_2, is straightforward: $\boldsymbol{H} = \frac{1}{2}(k_1 + k_2)$ and $\boldsymbol{K} = k_1 k_2$; note that k_1 and k_2 are also called principal curvatures. At this point, we want to make objection for using only $\boldsymbol{H}$ and $\boldsymbol{K}$ as features describing curvature. Their calculation is based on eigenvalues only omitting the information provided by the corresponding eigenvectors, i.e. the principal directions. Based on that, we propose to use the shape operator as our second feature which carries information about the shape of the surface in a local region and alleviating, in certain sense, the ambiguity introduced by the LARK features.

4.3.2 Multiple Kernel Learning for 3D Face and Emotion Recognition

Fusing the features, which were introduced in the previous sections, can be done in multitude of ways. In our work, we decided to use the SimpleMKL algorithm mainly due to two important reasons. First one is that the kernel methods are leader in the machine learning tasks and they are reliable tool, proven through the years. They are widely used in connection to Big Data and learning from large data sets. These methods can serve for achieving nonlinearity easily by embedding the data in a reproducing kernel Hilbert space (RKHS) [42]. Second, SimpleMKL induces sparsity in the weights of the kernels allowing for identification of the most relevant features (in the kernel space).

The core of MKL is learning convex combination of selected kernel matrices. Furthermore, the weights should be positive so can the resulting kernel matrix positive definite, which is the requirement for each kernel function for each couple of data samples:

$$K\left(x, x'\right) = \sum_{m=1}^{M} w_m K_m\left(x, x'\right), \; w_m \geq 0, \; \sum_{m=1}^{M} w_m = 1 \qquad (4.19)$$

Using this expression and the representer theorem [42], a function in RKHS induced by (4.19) is

$$f\left(\cdot\right) = \sum_{i=1}^{N} \sum_{m=1}^{M} \alpha_i w_m K_m\left(\cdot, x_i\right) + \beta, \qquad (4.20)$$

hence, the goal of MKL is to solve simultaneously for α_i, w_m and β. The difficulty solving the problem of MKL is imposed by the non-smoothness of the objective function, because the space of the kernels is a cone as can be seen in (4.19).

The formulation of MKL as convex optimization problem is the following [38]:

$$\min_{\{f_m\}, b, \xi, d} \frac{1}{2} \sum_m \frac{1}{d_m} \|f_m\|_{\mathcal{H}_m}^2 + C \sum_i \xi_i \qquad (4.21)$$

$$\text{s.t.} \ \ y_i \left(\sum_i f_m (x_i) + b \right) \geq 1 - \xi_i, \ \forall i$$

$$\xi_i \geq 0, \ \forall i$$

$$\sum_m d_m = 1, \ \ d_m \geq 1, \ \forall m$$

where $\|\cdot\|_{\mathcal{H}_m}$ is the respective RKHS norm. Minimizing over d_m with positive constraints is similar to minimizing its ℓ_1 norm and hence the sparsity introduced by SimpleMKL. Instead of solving (4.21), the authors of SimpleMKL [38] propose to solve the constrained optimization problem:

$$\min_d J (d) \ such that \sum_m d_m = 1, \ \ d_m \geq 1, \ \forall m \tag{4.22}$$

where $J(d)$ is equal to

$$\begin{cases} \min_{\{f_m\},b,\xi,d} \frac{1}{2} \sum_m \frac{1}{d_m} \|f_m\|_{\mathcal{H}_m}^2 + C \sum_i \xi_i \\ \text{s.t.} \ \ y_i \left(\sum_i f_m (x_i) + b \right) \geq 1 - \xi_i, \ \forall i \\ \xi_i \geq 0, \ \forall i \end{cases} \tag{4.23}$$

Both optimization problems are solved iteratively until stopping criteria are met. The first one (4.22) can be solved using the simplex method, while the second one is solved by reduced gradient algorithm. Detailed information on SimpleMKL can be found in [38].

4.3.3 3D Face Recognition

For this framework, in order to fully exploit the advantages of the MKL, the two previously described features – surface normal and LARK – are "kernelized". This is done by selecting appropriate kernel functions for each feature. In our case, it is important for the kernel functions to have a free parameter which allows for building multiple kernels. Since the surface normals are unit length vectors, we use the polynomial kernel which is more appropriate for normalized data since it exploits a dot product between features:

$$k (x_1, x_2) = \left(\alpha x_1^T x_2 + \beta \right)^d. \tag{4.24}$$

For the LARK features, we selected the Gaussian kernel:

$$k (x_1, x_2) = \exp \left(-\gamma \|x_1 - x_2\|^2 \right). \tag{4.25}$$

The parameter d is fixed, while the parameters α, β and γ are varied in order to build multiple kernels. Using the procedures described above, the proposed framework is defined as follows: (1) extract relevant face data, filter and resample; (2) extract surface normals and LARK features; (3) build kernel matrices using polynomial and Gaussian kernels; and (4) learn (apply) discriminant function using SimpleMKL algorithm.

We performed tests of the proposed approach using the "SHape REtrieval Contest 2008: 3D Face Scans" (SHREC) database [43]. All face-scanned subjects in this database are from Caucasian people, scanned with Minolta Vi-700 laser range scanner.

There are seven scans per subject, two are with neutral expression and the rest are with expression. Thus, there are totally 61 subjects with 7 scans per subject (427 scans), namely two "frontal", one "look-up", one "look-down", one "smile", one "laugh" and one "random expression".

SHREC scans are normalized for pose variations, and a simple hole filling algorithm is applied. Moreover, the tip of the nose is centred at the origin of the 3D coordinate system. This kind of normalization reduces the dependency of irrelevant variations and emphasizes on the actual 3D face recognition. A sample of the database is presented in Figure 4.14.

To evaluate the proposed framework, two experiments have been run on with the SHREC data set. For training, two face scans at random are selected in Experiment 1 and three face scans in Experiment 2, and the rest are used for testing.

The parameter selection for each step of the framework is as follows: (a) the radius R_s for face extraction is set to 100 mm; (b) the size of the median filter window N_{med} is set to 5; (c) the sizes of the PCA analysis window N_{PCA} and the LARK analysis window N_{LARK} are set to 7; (d) both parameters for the polynomial kernel vary in the range –10 to 10 with Step 1, and the power d is set to 4; and (e) the parameter γ for the Gaussian kernel varies in the range 0.5 to 5 with steps of 0.5. The experimental results are presented in Table 4.1 and compared with [44].

Figure 4.14   Sample from the 3D face scans of SHape REtrieval Contest 2008

Table 4.1 Experimental results

Experiment	Recognition Rate (%)
Experiment 1	92.73
Experiment 2	94.90
Result [44]	80.8

4.3.4 3D Facial Expression Recognition

Testing the credibility of the proposed algorithm is done on the BU-3DFE database [45]. It contains 100 subjects in approximately equal quantities, with age ranging from 18 years to 70 years and a variety of ethnic/racial ancestries, including White, Black, East Asian, Middle East Asian, Indian and Hispanic Latino. The subjects are annotated with 7 universal emotions: anger, contempt, disgust, fear, happy, sad and surprise. A sample for one subject from the BU-3DFE database is presented in Figure 4.15.

Initially, the 3D face scans are resampled on a common grid over the x, y plane. Afterwards, each face is aligned using the ICP algorithm [46] to the average face calculated using all samples in BU-3DFE.

LARK features are calculated over a window 7×7, with overlapping between consecutive windows of 2 pixels. Calculating the shape operator is performed in two steps. First, the average of the coefficients of the first and second fundamental forms is calculated in a window of size 3×3. Next, smoothing is performed by regularizing the singular values as described in [39], and the shape operator is calculated as in (4.18).

In this framework, we use the LARK and shape operator features. Building the kernel matrices for both features is performed using the standard Gaussian kernel $k\,(\mathrm{a}, \mathrm{b}) = \exp\left(-\gamma dist(\mathrm{a}, \mathrm{b})^2\right)$. For LARK features, $dist(*, *)$ is the χ^2 similarity, and for the shape operator, the Frobenius norm is used. The range of the coefficients for both kernels is from 0.1 to 10 with steps of 0.5.

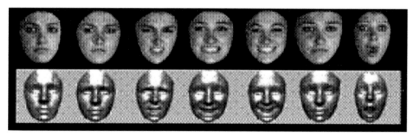

Figure 4.15 Sample face images and 3D scans from BU-3DFE database

Table 4.2   Confusion matrix

%	Anger	Disgust	Fear	Happy	Sad	Surprise
Anger	**80.1**	5.1	2.3	1.12	15.0	0.2
Disgust	5.9	**73.3**	7.12	4.8	3.1	4.2
Fear	4.2	7.9	**71.4**	8.2	4.14	3.87
Happy	1.1	1.25	5.88	**87.73**	1.89	1.53
Sad	9.5	4.46	8.34	3.66	**76.46**	1.35
Surprise	0.26	2.0	3.92	2.82	3.59	**90.5**

Table 4.3   Comparison with similar work

Author	%
Berretti et al. [48]	77.5
Gong et al. [49]	76.22
Li et al. [47]	80.14
Our approach	**79.91**

The experimental protocol is as follows: a subset of 60 subjects is selected from 100 which in turn is divided into 54 subjects for training and 6 for testing. This is repeated 100 times, and the averaged score is considered at the end. A similar procedure is proposed in [47]. The average confusion matrix for the 7 basic emotions is presented in Table 4.2.

Some emotions are more difficult to separate from others such as fear, disgust and sadness, because even in real-life situations, these emotions have very similar expressions.

The overall recognition rate of the algorithm is 79.91 %, and comparison with other algorithms is presented in Table 4.3. Note that the experimental protocol used in the referenced works is similar but not exact.

4.4 Conclusion and Comments

In this chapter, we have presented a framework using an active multicamera system in 3D space for human gaze tracking and face orientation determination combined with 3D face and emotion recognition based on multiple kernel learning. The framework is suitable for being part of assistive medical system for neurorehabilitation of patients with TBI or other type of brain damage. The recognition results of the proposed algorithms offer comparable performance in terms of facial emotional recognition rate (79.91%) in comparison with some of the state-of-the-art 3D facial expression recognition techniques.

It is difficult to do an exact comparison due to differences in the experimental protocol used in the referenced works as no standard testing protocol is available to compare between different 3D face recognition systems [57].

The goal is to follow remotely the progress of the patient when he or she is accomplishing some mental tasks, if the patient is paying attention, what is his emotional state or is he feeling fatigue. By using the power of 3D image processing, we can also incorporate this framework as being part of virtual reality (VR) medical systems. This hypothesis is supported by the current proliferation of VR-based telerehabilitation systems [50] and has enabled new paths for the development of multimodal scenarios supporting multisensory interaction. Such systems can be effective and motivating for rehabilitation therapies involving repetition and feedback which is very suitable for patients with brain damage [51]. In particular, there is evidence for the effectiveness of such approaches for the rehabilitation of upper limbs in patients with stroke [52, 53].

Additional experiments are needed with different set of features and different MKL algorithms in order to investigate the full applicability of the proposed framework. Also, we need to consider how to gather and use the 3D facial information from the Kinect sensor and to incorporate it fully in the proposed framework.

As it was stated in the introduction, the publicly available image data sets with different emotional states are captured in non-natural/controlled environment, so the developed algorithms for emotion recognition will not be as accurate when confronted in real-life situations, such as neurorehabilitation applications. Extensive data sets in real-life situation need to be gathered. Further improvements of this framework are needed for the purpose of neurorehabilitation to tackle the problem of the lack of rotational movement freedom causing problems for the facial and emotional recognition system because of the missing facial data.

Acknowledgment

This work was supported in part by the Grant Agreement No: 610658, eWALL: eWall for Active Long Living" of the EU Seventh Framework Programme.

References

[1] http://www.headinjuryctr-stl.org/statistics.html
[2] https://www.braintrauma.org/tbi-faqs/tbi-statistics/

[3] Report : Injuries in the European Union, Summary of injury statistics for the years 2008–2010, Issue 4 (http://ec.europa.eu/health/data_collection/docs/idb_report_2013_en.pdf)

[4] Roozenbeek B., Andrew I. R. Menon M. & D. K.: Changing patterns in the epidemiology of traumatic brain injury, Nature Reviews Neurology, vol. 9, pp. 231–236 (April 2013).

[5] Milders, M., Fuchs, S., & Crawford, J. R. Neuropsychological impairments and changes in emotional and social behaviour following severe traumatic brain injury. *Journal of Clinical & Experimental Neuropsychology*, 25, pp. 157–172, (2003).

[6] Williams H., Neuro-rehabilitation & Emotion, Centre for Clinical Neuropsychological Research (CCNR), School of Psychology University of Exeter.

[7] http://www.tbicare.eu/

[8] Perez-Marcos D., Solazzi M., Steptoe W., Oyekoya O., Frisoli A., Weyrich T., Steed A., Tecchia F., Slater M., Sanchez-Vives M. V.: A fully-immersive set-up for remote interaction and neurorehabilitation based on virtual body ownership. *Frontiersin Neurology* (2012) 3:110. doi: 10.3389/fneur.2012.00110

[9] Cranen, K., Drossaert, C. H., Brinkman, E. S., Braakman-Jansen, A. L., Ijzerman, M. J., and Vollenbroek-Hutten, M. M. (2011). An exploration of chronic pain patients' perceptions of home telerehabilitation services. *Health Expect* 2012 Dec 23;15(4):339–50.

[10] Golomb, M. R., Mcdonald, B. C., Warden, S. J., Yonkman, J., Saykin, A. J., Shirley, B., Huber, M., Rabin, B., Abdelbaky, M., Nwosu, M. E., Barkat-Masih, M., and Burdea, G. C. (2010). In-home virtual reality videogame telerehabilitation in adolescents with hemiplegic cerebral palsy. Arch. Phys. Med. Rehabil. 91, 1–8 e1.

[11] Gitlin LN, Vause Earland T. 2010. Dementia (Improving Quality of Life in Individuals with Dementia: The Role of Nonpharmacologic Approaches in Rehabilitation). In: JH Stone, M Blouin, editors. *International Encyclopedia of Rehabilitation.* Available online: http://cirrie.buffalo.edu/encyclopedia/en/article/28/

[12] Marker detector with opencv (Jun 2013), https://sites.google.com/site/playwithopencv/home/markerdetect

[13] Baluja, S., Pomerleau, D.: Non-intrusive gaze tracking using artificial neural networks. Tech. rep., Pittsburgh, PA, USA (1994).

[14] Bulling, A.,Ward, J., Gellersen, H., Troster, G.: Eye movement analysis for activityrecognition using electrooculography. Pattern Analysis

and Machine Intelligence, IEEE Transactions on 33(4), 741–753 (April 2011).

[15] Shan Gao, X., Hou, X. R., Tang, J., Cheng, H. F.: Complete solution classification forthe perspective-three-point problem. Pattern Analysis and Machine Intelligence, IEEE Transactions on 25(8), 930–943 (2003).

[16] Hansen, D., Hansen, J., Nielsen, M., Johansen, A., Stegmann, M.: Eye typing using markov and active appearance models. In: Applications of Computer Vision, 2002.(WACV 2002). Proceedings. Sixth IEEE Workshop on. pp. 132–136 (2002).

[17] Hansen, D., Ji, Q.: In the eye of the beholder: A survey of models for eyes and gaze. Pattern Analysis and Machine Intelligence, IEEE Transactions on 32(3), 478–500 (March 2010).

[18] Hutchinson, T., White, K. P., J., Martin, W. N., Reichert, K., Frey, L.: Human-computer interaction using eye-gaze input. Systems, Man and Cybernetics, IEEE Transactions on 19(6), 1527–1534 (Nov 1989).

[19] Ishikawa, T., Baker, S., Matthews, I., Kanade, T.: Passive driver gaze tracking with active appearance models. Tech. Rep. CMU-RI-TR-04-08, Robotics Institute, Pittsburgh, PA (February 2004).

[20] Jacob, R. J. K.: What you look at is what you get: Eye movement-based interaction techniques. In: Proceedings of the SIGCHI Conference on Human Factors in Computing Systems. pp. 11–18. CHI '90, ACM, New York, NY, USA (1990).

[21] Khosravi, M. H., Safabakhsh, R.: Human eye sclera detection and tracking usinga modified time-adaptive self-organizing map. Pattern Recognition 41(8), 2571–2593 (2008).

[22] Kumar, M., Garfinkel, T., Boneh, D., Winograd, T.: Reducing shoulder-surfing byusing gaze-based password entry. In: Proceedings of the 3rd Symposium on Usable Privacy and Security. pp. 13–19. SOUPS '07, ACM, New York, NY, USA (2007).

[23] Kumar, M., Paepcke, A., Winograd, T.: Eyepoint: Practical pointing and selection using gaze and keyboard. In: Proceedings of the SIGCHI Conference on Human Factors in Computing Systems. pp. 421–430. CHI '07, ACM, New York, NY, USA (2007).

[24] Kumar, M., Winograd, T.: Gaze-enhanced scrolling techniques. In: Proceedings of the 20th Annual ACM Symposium on User Interface Software and Technology. pp. 213–216. UIST '07, ACM, New York, NY, USA (2007).

[25] Spong, M., Hutchinson, S., Vidyasagar, M.: Robot Modeling and Control. Wiley (2006).

[26] Tan, K. H., Kriegman, D., Ahuja, N.: Appearance-based eye gaze estimation. In: Applications of Computer Vision, 2002. (WACV 2002). Proceedings. Sixth IEEEWorkshop on. pp. 191–195 (2002).

[27] Viola, P., Jones, M. J.: Robust real-time face detection. Int. J. Comput. Vision 57(2), 137–154 (May 2004).

[28] Williams, O., Blake, A., Cipolla, R.: Sparse and semi-supervised visual mappingwith the s³gp. In: Proceedings of the 2006 IEEE Computer Society Conference on Computer Vision and Pattern Recognition - Volume 1. pp. 230–237. CVPR '06, IEEE Computer Society, Washington, DC, USA (2006).

[29] Xiong, X., de la Torre, F.: Supervised descent method and its applications to facealignment. In: Computer Vision and Pattern Recognition (CVPR), 2013 IEEE Conference on. pp. 532–539 (June 2013).

[30] Eason G., Noble B., Sneddon I. N., Three-Dimensional Face Recognition in the Presence of Facial Expressions: An Annotated Deformable Model Approach, *Journal of IEEE Transactions on Pattern Analysis and Machine Intelligence*, Volume 29 Issue 4, April 2007, pp. 640–649.

[31] Bowyer K., Chang K., Flynn P., A survey of approaches and challenges in 3D and multi-modal 3D + 2D face recognition, *CVIU*, vol. 101, no. 1, pp. 1–15, 2006.

[32] Scheenstra A., Ruifrok A., Veltkamp R. C., A Survey of 3D Face Recognition Methods, in *AVBPA*, pp. 891–899, 2005.

[33] Campbel R., Flynn P., A Survey of Free-Form Object Representation and Recognition Techniques, *Computer Vision and Image Understanding*, 81, pp. 166–210, 2001.

[34] Hallinan P., Gordon G., Yuille A., Giblin P., Mumford D., *Two and Three-dimensional patterns of the face*, Ed. A. K. Peters, 1999.

[35] Moreno A. B., Snchez, J. F. Vlez, F. J. Daz, Face recognition using 3D surface-extracted descriptors, Irish Machine Vision and Image Processing Conference, 2003.

[36] Gökberk B., Dutağacl H., Ula_ A., Akarun L., Sankur B., Representation Plurality and Fusion for 3D Face Recognition, IEEE Transactions on Systems, Man, and Cybernetics-Part B: Cybernetics, Volume: 38, Issue: 1, pp. 155–173, 2008.

[37] Seo H. J., Milanfar P., Face Verification Using the LARK Face Representation, IEEE Transactions on Information Forensics and Security, 2011.

[38] Rakotomamonjy A., Bach F. R., Canu S., Grandvalet Y., Simple MKL, Journal of Machine Learning Research 9, 2008.

[39] Hae Jong Seo, Milanfar, P., Face Verification Using the LARK Representation, Information Forensics and Security, IEEE Transactions on (Volume: 6, Issue: 4), pp. 1275–1286, 2011.

[40] Abbena E., Salamon S., Gray A., Modern Differential Geometry of Curves and Surfaces with Mathematica, 3rd ed., Chapman and Hall/CRC, 2006.

[41] Snyder J., Deriving the Weingarten Map, Microsoft Research, 7/1/2011.

[42] Schlkopf B., Smola A., Learning with Kernels, MIT Press, 2001.

[43] TerHaar F. B., Daoudi M., Veltkamp R. C., SHapeREtrieval contest 2008: 3D face scans, IEEE International Conference on Shape Modeling and Applications, 2008, pp. 225–226.

[44] Nair P., Cavallaro A., SHREC08 Entry: Registration and Retrieval of 3D Faces using a Point Distribution Model, Shape Modeling and Applications 2008, IEEE International Conference on, 2008, pp. 257–258.

[45] Yin L.; Wei X., Sun Y., Wang J., Rosato M. J., A 3D facial expression database for facial behavior research, FGR, pp. 211–216, 2006.

[46] Besl P. J., McKay N. D., A method for registration of 3-D shapes, IEEE PAMI, vol.14, no.2, pp. 239–256, Feb 1992.

[47] Li H., Chen L., Huang D., Wang Y., Morvan J., 3D facial expression recognition via multiple kernel learning of Multi-Scale Local Normal Patterns, ICPR, pp. 2577–2580, 2012.

[48] Berretti S., Bimbo A. D., Pala P., Amor B. B., Daoudi M., A Set of Selected SIFT Features for 3D Facial Expression Recognition, ICPR, pp. 4125–4128, 2010.

[49] Gong B., Wang Y., Liu J., Tang X., Automatic facial expression recognition on a single 3D face by exploring shape deformation, In ACM Multimedia, 2009.

[50] Brochard, S., Robertson, J., Medee, B., and Remy-Neris, O. (2010). What's new in new technologies for upper extremity rehabilitation? Curr. Opin. Neurol. 23, 683–687.

[51] Holden, M. K. (2005). Virtual environments for motor rehabilitation: review. Cyberpsychol. Behav. 8, 187–211; discussion 212–189.

[52] Levin, M. F. (2011). Can virtual reality offer enriched environments for rehabilitation? Expert Rev. Neurother. 11, 153–155.

[53] Lucca, L. F. (2009). Virtual reality and motor rehabilitation of the upper limb after stroke: a generation of progress? J. Rehabil. Med. 41, 1003–1100.

[54] Grother P., Ngan M., Face Recognition Vendor Test (FRVT), Performance of Face Identification Algorithms, NIST Interagency Report 8009,

Information Access Division, National Institute of Standards and Technology, 26.05.2014.

[55] Williams M., Better Face-Recognition Software, MIT technology review, 30 May 2007.

[56] http://www.numerical-tours.com/matlab/fastmarching_4bis_geodesicx_mesh/

[57] Zaeri N., 3D Face Recognition, in *New Approaches to Characterization and Recognition of Faces,* edited by Peter Corcoran, ISBN 978-953-307-515-0, InTech, August 01, 2011.

[58] Panev S., Petrov P., Boumbarov O., Tonchev K., Human gaze tracking in 3D space with an active multi-camera system, Intelligent Data Acquisition and Advanced Computing Systems (IDAACS), 12-14.09.2013 Berlin Germany, Vol. 1, pp. 419–424.

Biographies

K. Tonchev is a researcher at the Faculty of Telecommunications, Technical University of Sofia, Bulgaria. His main interests are Kernel-based Support Vector Machines, age and gender recognition, 3D face models, computer vision. Mr. Tonchev is currently finishing his PhD degree. He has participated in several scientific projects both national and international for the development of reliable face and emotion recognition methods both in 2D and 3D. With collaboration with other authors he has published several papers in conferences such as WSEAS, IDAACS and others. He is an IEEE member.

S. Panev is a researcher at the Department of Radio communications and Video Technologies, Faculty of Telecommunications, Technical University of Sofia, Bulgaria. He finished his PhD in 2014. His main areas of expertise are pupil, gaze tracking, face orientation determination, modeling of active multi-camera systems for 2D and 3D face processing, computer vision etc. Mr. Panev is laureate of Fulbright scholarship for 2014 and he will work at the Computer Vision Laboratory at the Carnegie Mellon University, Pittsburgh, USA. He has participated in several scientific projects both national and international for the development of reliable face and emotion recognition methods both in 2D and 3D.

Assist. Prof. A. Manolova is with the Faculty of Telecommunications at the Technical University of Sofia (TU-Sofia), Bulgaria. She finished her PhD conjointly between the TU-Sofia and the Joseph Fourier University, Grenoble, France. Her domains of interest are Pattern Recognition, Computer Vision, Statistics, Image and Video processing, Processing of multispectral and hyper spectral images. Ms Manolova has participated in several scientific projects both national and international. She is laureate of Fulbright scholarship for 2013 working at the Computer Vision Laboratory at the Little Rock University, Arkansas, USA on a project concerning recognition of human emotions. She speaks fluently French and English.

N. Neshov is assistant professor and young researcher with the Faculty of Telecommunications at the Technical University of Sofia (TU-Sofia), Bulgaria. He finished his PhD in 2014. His domains of interest are content based image retrieval systems, face and emotion recognition, 3D image processing and machine learning. Mr. Neshov has participated in several scientific projects both national and international and currently he is teaching courses for Bachelor and Master degree students at the Faculty of Telecommunications. Mr. Neshov is also involved in developing an online educational system for the students at the TU-Sofia.

Prof. O. Boumbarov is the leader of the research laboratory "Electronic systems for visual information" at the Faculty of Telecommunications, Technical University of Sofia, Bulgaria. He is a renowned researcher in the field of computer vision in Bulgaria and under his supervision many nation and international projects were successfully concluded. His main scientific interests are 3D face image processing, face and emotion recognition, neural networks, EEG and ECG signal processing, biometrics etc. He is a longtime professor at the Technical University, teaching courses for Bachelor and Master degree students. During his career he has published more than 300 scientific papers.

Professor V. Poulkov PhD, has received his MSc and PhD degrees at the Technical University of Sofia, Bulgaria. He has more than 35 years of teaching and research experience in the field of telecommunications. The major fields of scientific interest are in the field of information transmission theory, modulation and coding. His has expertize in the field of interference suppression, power control and resourse management for next generation telecommunications networks. Currently he is Dean of the Faculty of Telecommunications, Head of "Teleinfrastructure" and "Electromagnetic Compatibility of Communication Systems" R&D Laboratories at the Technical University of Sofia, Thematic Area Leader "Optimal Resource and Embedded ICT" at the Center of Teleinfrastructure - Aalborg University, Chairman of Bulgarian Telecommunications Cluster, Senior IEEE Member, Member of CONASENSE society.

5

An Integrated Perspective for Future Widespread Integration of Neuro-motor Rehabilitation

Giulia Cisotto[1,2] and Silvano Pupolin[3]

[1]Keio Institute of Pure and Applied Sciences, Faculty of Science
and Technology, Keio University, Yokohama, Japan
[2]Integrative Brain Imaging Center, National Center of Neurology
and Psychiatry, Tokyo, Japan
[3]Department of Information Engineering, University of Padua, Padova, Italy
Corresponding author: Silvano Pupolin <silvano.pupolin@unipd.it>

5.1 Introduction

In recent years, Neuro-motor Rehabilitation had a push to use information and communication technology (ICT) in order to enhance the natural recovery after a brain injury. From start onwards, one of the key ideas was to understand and to use the brain activities related to movements. This development began from the papers by Vidal [1], Farewell and Donchin [2] and Birbaumer [3] which proposed to use electroencephalogram (EEG) signals to drive prosthesis. In particular, Birbaumer used these signals to provide completely paralysed persons a contact with their caregivers, again. This last idea was extended to capture from EEG signal features related to the limb movements in order to give real-time feedback to the patient. Some preliminary results [4] have shown that by designing appropriated protocols for physical exercises, this system improves patient performance in recovery motor functions with respect to the standard Neuro-motor Rehabilitation method.

Two other ICT-driven systems could be employed to give a real-time feedback to the patient and move a robot to perform a movement: they are the virtual reality (VR)- and electromyography (EMG)-based biofeedback platforms.

Neuro-Rehabilitation with Brain Interface, 89–120.

In this chapter, we analyse the current status of physical recovery after brain injury, the ICT rehabilitation devices available nowadays and their typical use. Since the early stage of application of the Neuro-motor Rehabilitation, the available ICT is rather complex and it still requires specialized operators, so its use is limited to hospitals or rehabilitation centres.

Since Neuro-motor Rehabilitation requires a long-term training and advance instrumentations, this could be available only at specialized centres, where the high costs of those devices and their maintenance are not fully sustainable by national health organizations. Nevertheless, if we could help brain-injured people through technology, we may lead them to regain some independence in accomplishing the activities of their daily lives reducing, at the same time, their need to visit specialized centres. In this way, we may achieve a twofold advantage: on the one hand, patients could be back to their work, and on the other hand, the number of full-time caregivers could be significantly reduced. In order to achieve this scope, we should substantially decrease the cost of Neuro-motor Rehabilitation: one promising way is by using ICT rehabilitation instruments that could be operated by the patients themselves at their home. This could appear rather hard. Nevertheless, it is actually not an impossible task. Let us think how we are currently using complex systems in a simple way: smartphones are the most exemplifying instance where extremely complicated technology could be used by everyone, intuitively. This could represent the same perspective for the design and operations of the rehabilitation in future.

From this point of view, technical and clinical guidelines are highly needed. First of all, we need simple interfaces for the device operation; secondly, it has to be guaranteed the correct system functioning even at the patient's home where no specialized operators are present. Finally, the ICT activities should be remotely controlled by a rehabilitation centre where specific software should be developed to extract interesting features to characterize the patient status in every single time. Similarly, the software should be used to translate the extracted features into useful commands delivered to the devices supporting the patient in his/her home.

Not least, the instructions how to use the instruments should be clearly explained to the patient, e.g. by an introductory video, in order to let the devices operate at the most effective way.

Before the installation of the devices at home, a careful design of the most robust algorithms to be implemented is totally required and it probably needs to be accomplished both at clinical institutes and at private homes.

A major aspect in the study focuses, on the other hand, on the effects of ICT-driven feedback on the patient: benefit expectation and the definition of rehabilitative ICT-based programmes require a deep knowledge of the neurological mechanisms behind the patient movement. From the engineering viewpoint, an accurate and detailed design of an effective and easy use of the system is the most important issue. For example, the best sensors to capture the signals have to be chosen in terms of novel, robust, easy-to-use and low-cost system. In order to succeed in this aim, a multidisciplinary team composed of several different specialists, e.g. neurologists, psychologists, therapists, engineers and market experts, is definitely needed.

Indeed, if we want to spread the use of these devices, they must be characterized by low costs and simplicity of usage. Therefore, a solution that used either magnetic resonance imaging (MRI) or magnetoencephalogram (MEG) cannot be really taken into account due to their high costs and complexity. Similarly, EEG could unlikely represent a reliable solution since each of its sensors needs to be appropriately mounted on the head of the person to establish an accurate electrical contact with the skin and allow a good signal-to-noise ratio during acquisitions. A new system to capture brain signals which seems to be suitable for home usage is based on near-infrared spectroscopy (NIRS) and its functional version (fNIRS). In Figure 5.1, examples of an MRI scanner, and MEG, EEG and NIRS equipments are shown.

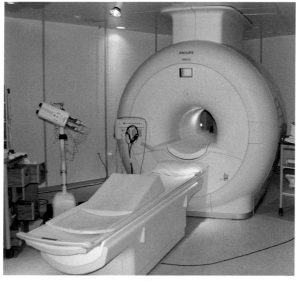

(a)

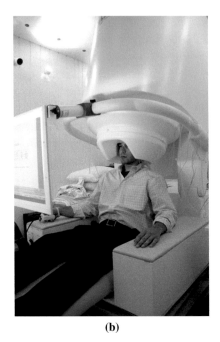

(b)

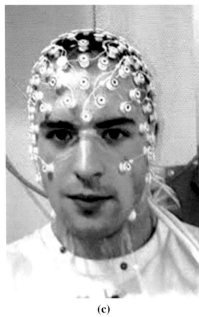

(c)

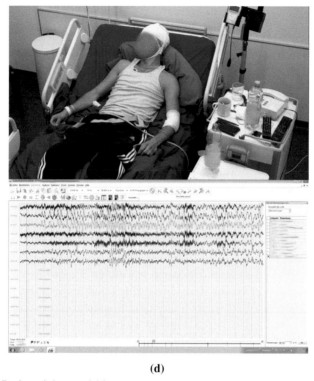

(d)

Figure 5.1 Brain activity acquisition systems: a) MRI scanner, b) MEG, c) NIRS and d) EEG (Courtesy of MicromedS.p.A, Mogliano Veneto, Italy)

An overview of the today's knowledge on ICT for rehabilitation is outlined as well as the perspective of its development during the next decade: "full-scale integration" and "home rehabilitation" are keywords in this context.

This chapter is organized as follows: Section 5.2 gives a look at the motor diseases and the rehabilitation techniques already available, while Section 5.3 illustrates the two main rehabilitation approaches. Then, Section 5.4 deals with the key principles of modern rehabilitation, and Section 5.5 is devoted to the optimization of these techniques. Finally, Section 5.6 sketches an idea for future perspective for rehabilitation, and Section 5.7 concludes the chapter.

5.2 Overview of Motor Diseases and Their Rehabilitation

Neurorehabilitation is the field of medicine which aims at achieving the recovery of functions, e.g. motor and cognitive, lost or partially damaged either after a traumatic injury or due to a degenerative disease.

Common causes of sudden function loss are traumatic brain and spinal cord injuries (TBI and SCI) mainly due to falls and road accidents, heart and cardiovascular diseases such as stroke and progressive neuropathologies as Parkinson's disease (PD), Alzheimer's and amyotrophic lateral sclerosis (ALS).

Among others, the World Health Organization (WHO) estimated that a stroke has most wide and significant impact on the population. Main risk factors are as follows: being of age, tobacco use, harmful use of alcohol, physical inactivity and unhealthy diet [5]. From the latter list, a wide coincidence of stroke in the global health of the most developed countries is easily understood.

Besides, even though stroke is the most common cause of mortality, it represents also one of the main causes of disability in the world. Indeed, survivors from stroke could survive for many years, but in 60% of the cases, they suffer from long-term functional impairments, where cognitive abilities, speech and motor functions are the most frequently impaired. This usually leads to a reduced independence and self-determination in the accomplishment of the activities of the daily life (ADL) and to situation in which caregivers are often needed 24 hours a day to help patients during their life.

Loss or severe damages in the sensory and motor functions probably represent the most disabling consequences of a stroke or other kinds of cerebrovascular disease (CVD): the incapacity of reaching objects, holding them, moving limbs and gait highly impact on the life of the patients and their families.

Not only CVDs seriously affect the daily life of more and more people at global scale, but also spinal cord injuries, tetraplegia, hemiplegia, ALS, PD and many other neuropathologies lead to impairments on the sensorimotor system that, in turn, make even the most elementary movements a difficult task.

The neurorehabilitation, i.e. sensorimotor rehabilitation, focuses on the recovery, partial at least, of sensorimotor functions from the more rough movements of the proximity segments of the body, i.e. shoulder and arm, to the finer actions performed by the hands. Even a little improvement in this kind of activities could mean a step towards the regaining of independence in the daily life of a patient. Figure 5.2 shows a finger rehabilitation which uses a robot.

Many decades of clinical experience and an increasing scientific support have established the beneficial role of the physical training and the repetitive training for stroke survivors, especially in the first three months after the acute event. Nowadays, international guidelines for the best clinical practice [6]

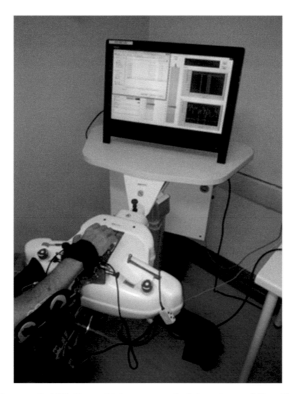

Figure 5.2 Finger rehabilitation which uses a robot (courtesy of San Camillo Hospital, Venice-Lido, Italy)

include several kinds of rehabilitative methods. Many of them are based on the repetition of motor functions with the affected arm; goal-directed and task-targeting exercises, occupational therapy, along with magnetotherapy and physical therapy, became milestones in the standard rehabilitative training after stroke [7, 8].

Although medicine and biology have recently obtained brilliant results, for instance stem cell studies made possible human retinal transplant [9] and spinal cord regeneration in non-human primates [10], a gold standard therapy for faster and more efficient recovery from stroke is still far away.

Many factors still influence the rehabilitative process; we mention here the physiological and psychological conditions, the environment and the efficacy of the treatment at the hospital where the patient was hospitalized suddenly after the stroke episode and the post-hospitalization training.

Nowadays, the hospitalization period in a rehabilitative institute lasts about three months; it is well known that the first three months after the stroke event (either haemorrhagic or ischemic) are most critical and cover the period in which the patient is more receptive, at the same time. Indeed, suddenly after the stroke event, spontaneous processes develop in the brain with the aim to restore or compensate for the functionalities compromised by the disease. All these kinds of processes, i.e. neural sprouting, neurogenesis, axon genesis [11, 12], dendritic branching and synaptogenesis, occur in the brain (revealed by animal studies) and represent the main mediators of the recovery [13–15]. It has been also proved that those spontaneous activities generally called *neuroplasticity* may be enhanced by repetitive training and goal-directed exercises.

Roughly speaking, neuroplasticity is neural system property that allows brain to exploit new paths – previously underused and redundant in the healthy brain – to restore abilities of the individual.

The general aim of the modern rehabilitation was to restore motor, cognitive and other functions taking advantage of this brain characteristic. It has to be mentioned that neuroplasticity could sometimes lead brain compensations instead of optimal recovery; therefore, suitable learning-based rehabilitative protocols have to be carefully designed.

For this reason, therefore, many basic studies (biology and electrophysiology) and recovery techniques have been establishing to clarify the principles for the best clinical practice.

Two main approaches are carried on in the modern rehabilitation field: the substitutive, compensatory or adaptive one and the restorative or facilitating one. With special reference to the Neuro-motor Rehabilitation, in the following, a general explanation of both of them will be provided and examples from the current clinical practice will be also given. Subsequently, an overview of the most advanced techniques will show the direction of the rehabilitation of the future.

5.3 Two Main Approaches to Rehabilitation

In dependence of the kind of disease, of the patient and of the guidelines of the rehabilitative institute, the team of medical doctors, physical therapists, neuropsychologists and technicians who are in charge of selecting the most suitable motor recovery programme for the patient has to take care of which kind of therapy they are going to administer. Just to underline this approach, it may be generally stated that for a typical stroke patient, a facilitating or restorative programme is offered to him/her immediately after the accident. Typically, physical therapies such as magnetotherapy, laser therapy, massage therapy and occupational therapy, along with a pharmacological treatment, are given within the first period of recovery, i.e. the first three months. Then, prognostic indicators, both qualitative (based on experience) and quantitative (based on physiological examinations), can give a hint about the most suitable rehabilitative training to administer in the following. Namely, if prognostic indexes predict a significant recovery, the restorative programme could be carried on with introduction of new repetitive, high-intensity, goal-directed exercises and their adaptation to the daily life.

Moreover, several other rehabilitative techniques are also available to enhance spontaneous restoration: biofeedback, above all, has a primary role.

As the term suggests, *biofeedback* means that a biological response is used as a feedback to the patient – i.e. (s)he is made aware about the intensity or frequency of that response – and (s)he has, in turn, to correct his/her behaviour or motor response in relation to it. Common types of feedback are EMG feedback, in which the level of activation of a target muscle is controlled by the subject in a closed-loop scheme [16]. This is especially useful in case of dystonic patients, who cannot control agonist and antagonist muscles separately anymore, and a feedback of their individual activation could make the patient aware of the contraction level in each of them (see Figure 5.3), with the possibility to adjust it and receive immediate feedback about the consequence of its reduction.

For stroke patients, instead, a helpful biofeedback technique is to administer a weak electrical stimulation (functional electrical stimulation (FES)) [18] on a target muscle, for example the extensor carpi radialis (ECR), in dependence on the residual EMG activity of the patient, in order to help him/her to restore an activity like reaching, grasping and holding. A good prototype of this rehabilitative approach is the one developed by [19].

Other kinds of therapies were conceived: in order to force recovery of an affected arm or hand, the healthy limb could be blocked for a period and the patient is so obliged to use the injured one for a period of 2–3 weeks (up to 6 hours per day). This technique, called the constraint-induced movement therapy (CIMT) [20], was shown to be highly effective in many stroke patients, with better results when performed within the first period (3–9 months) post-injury (see Figure 5.4).

On the contrary, during bilateral training, the subject is asked to perform identical activities with both his/her limbs simultaneously [21].

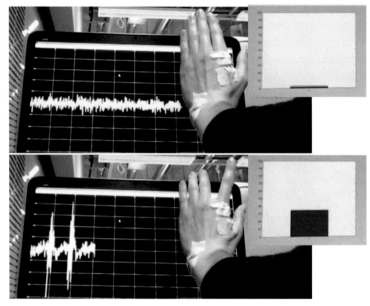

Figure 5.3   EMG biofeedback therapy. The EMG activity acquired by EMG surface electrodes is shown to the person, translated into the height of a vertical bar.

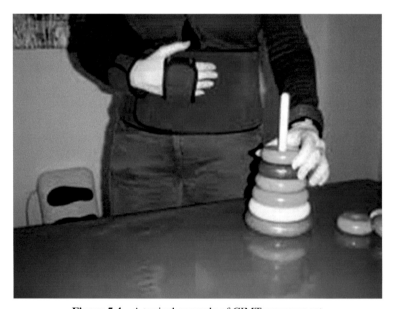

Figure 5.4   A typical example of CIMT arrangement

Taking advantage of the healthy arm or hand, mirror therapy [22] is an interesting technique that consists of performing a series of movements with the unaffected hand/arm without looking at it but, on the opposite, watching at its motion reflected on a mirror; surprisingly, it gives the illusion to move the affected hand/arm and this acts as a new afferent sensory input for the brain hemisphere injured by the stroke and was found effective for many stroke patients.

As Cochrane's surveys established, CIMT [23] as well as bilateral training [24] did not provide enough evidence to ascribe them into the best clinical practice guidelines, but occasionally, patient by patient, they could be applied, and thanks to them, the patient can achieve significant improvements.

Recently, enriched environments of VR [25] were also widely used to reinforce and enhance the spontaneous recovery processes via augmented proprioceptive, acoustic and visual feedback given by the system (see Figure 5.5). For example, a virtual kitchen could serve to train in a real-life situation reaching and holding tasks already trained with the physical therapist in the more controlled setting of a gym. At the same time, VRs can give visual and auditory warnings or stimuli, so that well-performed movements are rewarded.

When patients are not able to move at all due to the spasticity caused by the stroke, another frequently employed treatment is motor imagery (MI) [26]. This activity consists of imaging themself performing a specific movement, without any overt output. MI could be administered either in an *implicit* (the indirect consequence of the task accomplishment produces the movement) or in an *explicit* way (a usual task is to image the way to grasp

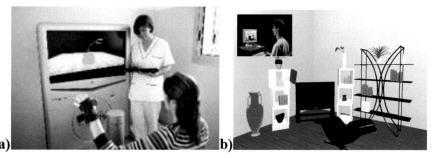

Figure 5.5 a) One of the VR systems at IRCCS San Camillo Hospital in Lido of Venice (Italy); b) an enriched environment of a typical VR system (Courtesy of Khymeia SRL, Padua, Italy).

an object), and more importantly, the most efficient type of MI for the Neuro-motor Rehabilitation purpose is the kinesthetic one. In the latter, the subject images to move his/her limb or segment as (s)he would actually perform – *first person perspective*. This mental activity was shown to enhance the afferent proprioceptive input from the periphery to the central nervous system more than in case of a *third person* MI, where the subject images to move the limb as (s)he was observing the movement from an external perspective, instead [7, 26].

The MI strategy is becoming increasingly attractive because it can be also embedded in a so-called brain–computer interface (BCI) or brain–machine interface (BMI) system [27–30]. BCI and BMI itself is a recent communication platform that allows direct interactions between one's brain and a computer or another external device. As any other communication system, it needs its own hardware and protocols. It means that a device is needed to acquire brain responses, e.g. an EEG, MRI, MEG and NIRS, as shown in Figure 5.1 above. Then, some features of the brain activity related to the task being performed by the patient have to be identified and extracted in real time by the computer that is connected with the patient's brain. Next, a *training* period or a training data set is required for the two interacting systems to share information and optimize the communication. For example, if a motor-disabled patient – let us say a patient with tetraplegia – wants to move a robotic arm, thanks to a microarray of electrodes implanted on the surface of his/her cerebral cortex, (s)he has to learn how to voluntarily modulate the specific brain features related to the movement and the computer (commanding the robotic arm) has to detect those voluntary changes. This is possible [31, 32], but in turn, an intelligent, automatic and adaptive detection algorithm has to be available on the computer. That allows the latter to discriminate physiological and task-related activities from different sources' activities and to adjust the robotic arm's motion parameters to the highly variable human brain response. The protocol in this particular application, then, is represented by the set of instructions the patient needs to learn in order to take advantage of the robotic arm and properly use it. On the other hand, the timing and real-time constraints require that the computer has to correctly process the physiological information and transform it into suitable robotic commands for the robotic device (see Figure 5.6).

Besides BCI, explicit kinesthetic MI can be used with stroke survivors who do not have any muscle activity. In this case, FES-based feedback of MI of grasping, holding or reaching tasks could lead the patients to develop – after a quite long and intense training – a minimum muscular activity that can

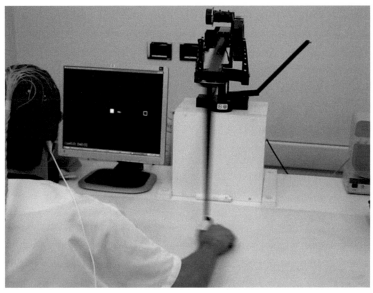

Figure 5.6 A BCI system for Neuro-motor Rehabilitation of stroke patients at IRCCS San Camillo Hospital in Lido of Venice (Italy)

be, subsequently, revealed by a surface EMG and used to drive a FES-based orthosis that helps the patient to perform the complete movement. This was proved to be effective for this kind of patients that, after such experimental treatment, could perform larger movements [19].

In other cases, MI could be related to simple visual or acoustic feedbacks that serve, basically, to investigate neurophysiological processes of the sensorimotor system. For instance, MI is of fundamental importance to discriminate the involvement of sensory and proprioceptive afferents coming from the periphery (e.g. effectors such as hands and muscles) compared to an actual movement [33]. When a person does not actually accomplish the movement, the only information (s)he can form in the brain is related to the planning phase of the motor action and the *memory* of past movements. Each of these components of the movement is widely studied by an expanding branch of neuroscience called *computational neuroscience* [34] that treats the brain as a control model where input, output and model states are black boxes whose characterization could be only guessed via evidences provided by behavioural aspects of the individual in response to particular tasks suitably designed to obtain specific information.

As said before, careful design of suitable protocols to shed light on the desired *brain block* has to be implemented. In order to do so, two main principles can be exploited: learning and neuroplasticity abilities of the brains and the possibility to establish a conditioning of its physiological activity by providing specific external stimuli.

5.4 Rehabilitative Basic Principles

A major role in the rehabilitation field, especially in the restorative or facilitating approach, is played as mentioned above by learning processes and conditioning techniques. Every rehabilitative technique is based on the capacity of each person to modify – voluntarily or not – his/her nervous system activity and, consequently, the motor output, i.e. behaviour.

When a CVD occurs in a cortical or subcortical region of the brain that is in charge of the sensorimotor activities of the individual, the latter most likely loses, at least partially, sensory and motor functions of the contralateral limbs [35]. Therefore, the patient became – in a sudden – impaired on how to use his/her hemibody. However, sensorimotor capabilities come from a very long experience-based training that each of us had developed from the very early stages of the childhood, during an intense *learning* period.

After a CVD, however, the patient could not rely anymore on the intact sensorimotor-related brain resources, i.e. neural assemblies, since they were disrupted or largely damaged by the injury.

Nevertheless, thanks to neuroplasticity, new resources – in terms of new neural pathways via new axons, new synapsis and the strengthening of previously redundant connections – could be exploited to compensate for the loss within and in the neighbourhood of the injured stroke area with a consequent and usually significant recovery of motor functions.

The main aim of the restorative Neuro-motor Rehabilitation was thus to induce or *facilitate* patients to *re*learn such functions without the usual neural resources exploited in the past of their life.

Many examples of (re)learning techniques had been already exploited in the clinical practice mostly without a complete awareness of that. Nowadays, computational neuroscience and other fields of neuroscience are largely investigating the learning procedures and the associated modifications in the anatomical neural structures in order to examine and discriminate the basic principles underlying the different kinds of learning; for example, reward-based learning and error-based learning are only a few. In future, each patient could be addressed to the optimal learning procedure depending on his/her state, severity of remaining impairment and response to different treatments in order to achieve the largest and fastest recovery.

It was shown for the first time by Fetz in [36] that stressing one neural path several times in response to a task requirement could form a new stable and reliable neural connection and, further, a new neural path passing through a number of neurons involved in the same task. This kind of *conditioned response* actuated by the neural assembly was named as *operant learning* or *operant conditioning*, in accordance with the definition of operant conditioning given by Skinner to more general learning processes in 1938 [37]; precisely, if a certain stimulus makes neuron A firing and producing an action potential that reaches neuron B and the latter fires in turn, and this scheme is repeated for several times, then once neuron A fires even in the absence of the stimulus, neuron B is stimulated and fires as well. Based on this principle, new restorative motor relearning techniques were established. One among others is the case of BCI. As explained in the previous sections, BCI allows establishing a real-time communication between a computer and a subject performing a specific task. In a rehabilitation perspective, BCI is used to identify and reinforce new neural pathways – like the one connecting neuron A and neuron B – giving a feedback (the stimulus of Fetz's operant conditioning) to the patient whenever (s)he performs the task exploiting the new alternative pathway [38].

To this scope, several laboratories are currently studying the most effective stimulating paradigms, brain signal processing and output control to optimize the modern BCI platforms. Anyway, BCI has been already shown to be an effective and promising tool to regain motor control in synergy with the standard physical therapy performed during the clinical practice [39].

Therefore, new therapies, new technologies and new clinical findings to improve sensory and motor recovery are being unrevealed nowadays: such tools could bring medicine to become more conservative of the remaining intact resources of the patients, and increasingly customized training programmes could make the latter recover as much as possible for each of them.

Besides the restorative or facilitating, the compensatory approach could also be used in rehabilitation: especially in the case in which, after the first three months post-stroke, no restoration of normal functions could be reasonably predicted, a substitutive approach seems to be more appropriate. For instance, if fine movements with the affected hand could not easily be recovered, a possible approach is to let the patient learn to use the other hand to accomplish to fine motor actions and the affected one acts as a support. This is one of the most common cases – knowing that the restoration of fine motility is a more difficult task than a rough control of the limbs. The patient frequently ends up to use the unaffected hand – no matter if it is the dominant or the other hand – to open the pots, the toothpaste, lock/unlock the door and so on, while the other hand sustains the pots, the toothpaste to prevent it to turn, etc.

In such circumstances, prosthesis is becoming increasingly popular: not only limited to arms, bionic eyes and artificial ears, but also robotic limbs [40] are nowadays common examples of technology successfully embedded in our body and useful to substitute lost functions without possibility of recovery.

BCI can thus be considered as a substitutive tool when it is used to control an external device, such as a hand, arm, leg or wheelchair, allowing the patient to regain, at least partially, his/her independence in the daily life. Preliminary successful examples of this are the robotic arm controlled by a woman suffering from tetraplegia implemented at Brown University in 2012 by Hochberg, Donoughe and collaborators [31] and a similar application developed in Pittsburgh by Collinger and colleagues last year [32], as mentioned earlier.

Those two successful experiences paved the way to a new era of the rehabilitation after sensorimotor injuries such as stroke, spinal cord injury or other severe brain damages. Nevertheless, a further experimental step on a larger population of patients is needed in order to make this complex platform a reliable and friendly device to be included in the clinical practice, first, and in the at-home rehabilitation environment in future. This would greatly help the recovery of sensorimotor functions of the patients, as well as the psychological condition of them and the care effort of caregivers and the patient's family.

Despite the selection of a facilitating or a substitutive rehabilitation programme, patients need a long training period to become familiar with the prosthesis and to slowly modify the brain activity used to accomplish to a movement. This familiarization or training activity could be only partially addressed at hospitals and rehabilitative institutes. Usually, after a first period of hospitalization, the patient has to move back to his/her home and continue the programme by his/herself with the help of private therapists and caregivers. It is easily understandable that very advanced rehabilitative programmes require a specialized and multidisciplinary team of clinicians and technicians. These services are far from common practice to be available at patient's home or in its neighbourhood. As a consequence, once people are dismissed from the hospital, they mostly stop their interrupting virtual reality exercises, EMG biofeedback, robotic and BCI training.

Several experimental programmes have been carried out recently [41–46] with the aim to create and test home-based rehabilitation platforms. Mostly, they include robotic training for improving arms or legs' functions after the stroke [41–43]. Other tentative activities have been developed to bring needed laboratory tools (especially brain signal acquisition, processing and output delivery), into mobile platforms even into smartphones [44]. These tools were originally used at hospitals or rehabilitative institutes, and in order to be effective, they need a control room (see Figure 5.7).

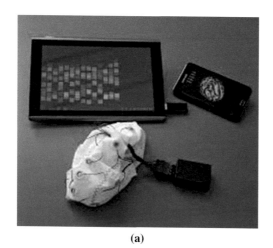

(a)

(b)

Figure 5.7 a) Smartphone brain scanner, and b) control room (Courtesy by Micromed SpA, Italy)

Not only stroke patients could benefit from these at-home rehabilitative systems to improve motor functions, but also other kinds of patients could benefit of them. Indeed, other platforms to recover non-motor functions have been already developed: for example, telemedicine can be employed to administer speech exercises to normalize the speech production of stroke patients suffering from aphasia as well as at-home BCI systems could allow ALS patients to communicate with their families and caregivers their primary needs.

Many features of those preliminary experimental platforms should be improved, and systems response and robustness should approach 100% reliability. In the subsequent sections, some comments and observations about the technical advancements of those technologies will be given, along with a final overview of the not-so-far future at-home programme.

5.5 Optimization of Current Rehabilitative Methods

Current advanced rehabilitative techniques make large use of ICT technologies applied to the physiological, neurological and medical knowledge. To design new rehabilitative protocols, long studies and live tests are needed to accurately design the system and make it robust against any parameter variability with time due to the clinical/psychological/physical status of the patient. Today, there are many research institutes/hospitals worldwide where the application of BCI/BMI to rehabilitation is done. There are still several problems to extend the use of such a system to small centres and at home. One of the main obstacles is related to the capture of brain signals by using EEG, which appears to be the best candidate to be used due to its relatively low cost and high temporal sensitivity. The main drawback of this system is its inherent difficulty to set correctly the EEG electrodes which require accurate positioning on the scalp and a good electrical contact between each electrode and the scalp itself. This requires an accurate scalp cleaning and the use of gel to get the electrical connection. This fact is probably the most annoying, time-consuming and difficult activity that should be done by the patient at home.

In recent years, a new measuring technique based on NIRS sensors is taking place and it began to be used together with the EEG. Its use requires further studies in order to design all the algorithms to drive properly the machines that are actually driven by the EEG signals. NIRS sensors appear to have more stable characteristics with time and a simpler and fast set-up than EEG sensors. Indeed, they could be located on a cap, similar to the one used for

EEG, but they do not have the necessity to create a good electric contact with the skull skin (see Figure 5.1).

Most of today's available rehabilitation machines are based on the use of bio-feedback which is derived from different sensors.

We divide the machines into two classes: 1) based on VR and 2) based on EMG signals. Machines using EEG signals are still in their infancy and are used in advanced research institutes/hospitals where accurate analysis of their performance to improve the rehabilitation process of the patients is still under investigation.

5.5.1 Virtual Reality

Virtual reality machines are based on the representation on a screen of the action that the patient has to implement. An example is the movement of the right hand in the space. The hand has sensors that identify its position and show it on the same screen where the action has been presented. At the beginning of the movement, the starting point of the action and the position of the hand on the screen are aligned. Then, performing the movement, the patient sees on the screen if (s)he is doing the correct path and the error. In this way, (s)he has a visual feedback of what (s)he is doing and of the error. This feedback is helpful to improve the arm movement and to do it in the best way.

To perform this exercise, it is necessary to define properly the exercises the patient has to do and to verify their efficacy.

Commercially available apparatus could be seen at web pages [47, 48].

In the last years, the development of 3D screens improved the efficacy of VR systems. Also, new techniques for generating the exercises have been developed; they are based on the use of games in order to stimulate the patient to perform the exercises. In this way, the therapy is less boring than the standard ones [49].

VR systems could also be used to perform speech rehabilitation by showing to the patient phrases and the related mouth movements and the replica as spoken by the patient itself (Figure 5.8).

Today's apparatus is composed of a 3D vision system, which typically uses a) one or more 3D Kinect^{TM} camera(s) used to capture either the face of the patient or the movement (s)he is doing; b) one or more PCs devoted to process the signals the system is capturing and extracting all needed features; c) a screen (possibly 3D); and d) a wideband communication device to be connected with the hospital/rehabilitation centre which is remotely taking care of the patient.

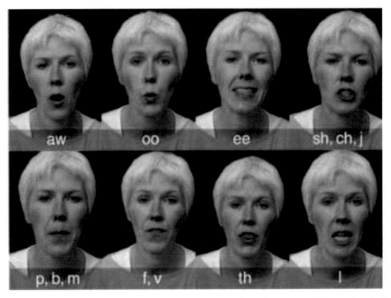

Figure 5.8   Virtual reality system for speech rehabilitation (Courtesy of Khymeia srl, Padua, Italy)

VR machines could be used also to support improvement in patient's cognitive ability and speech by proposing specific exercises prepared by neurologists. The add-on we need is a voice to text recognition system to serve as feedback for the speech rehabilitation and to give a feedback to the patient by translating the words pronounced by the patient in written text on the screen. In this way, the patient has an immediate feedback as the words pronounced are interpreted by the machine and is forced to a better pronunciation.

5.5.2 EMG Feedback

A proprioceptive feedback could be obtained by capturing the electric signals generated by the muscles when activated. We consider, for example, the movement of one hand. If we recognize the activation of the muscles to perform the movement by the EMG sensor, we could give a feedback to the patient that the muscles have been activated by several different means as activating either a buzzer, or a vibration, or a light signal. In this way, the patient has a feedback of the muscles' activation and (s)he could try to improve the force, if it is insufficient to perform the movement.

We could use EMG signals associated with a robotic device [50] where preassigned exercises are done in two different ways: i) the robot performs the arm movement, and the EMG measures the activation of muscles related to that movement and ii) the robot performs the arm movement by applying a force which depends on the EMG signals (see Figure 5.9). The larger the EMG signals, the lower the force applied by the robot. In this way, the patient has a direct proprioceptive feedback due to the fact that (s)he has to apply a force which, with the improvement of his/her reaction, is close to the one required to perform the movement without any assistance and iii) in case the EMG signals are below a threshold, the robot performs the movement, and through the EMG sensors, a small current is applied to activate the muscles (FES).

Looking at machines available to perform limb rehabilitation, we find two different solutions. The former devoted to upper and lower limbs, wrist and hand rehabilitation [51] which uses virtual reality and EMG neuromuscular stimulation, and the latter specialized to hand and finger rehabilitation [52] with both passive and active EMG stimulation.

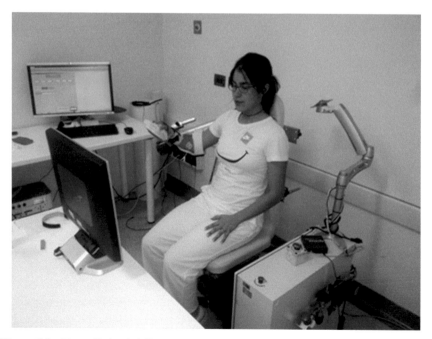

Figure 5.9 Upper limb rehabilitation by using EMG biofeedback (courtesy of San Camillo Hospital, Venice-Lido, Italy)

In conclusion, the today's available rehabilitation machines use VR and EMG signal for both feedback and stimulation. They show their efficacy, but some recent research results show that BCI/BMI could give a further improvement in the patient physical recovery of the lost functions. Then, future research results and ICT machines will add BCI/BMI techniques to the standard VR and EMG feedback to improve their efficacy.

5.6 Future Perspective

Moving from the actual situation what we foresee for the next decade is a wide-scale application of BMI to the rehabilitation protocols and their widespread use. There are still several problems to be solved related to the sensors used to detect brain signals. We are talking of non-invasive sensors as EEG, NIRS, MRI and functional magnetic resonance imaging (fMRI) or invasive sensors as electrocorticography (ECoG). The latter gives better and more precise signals than the former but needs an intracranial surgery which is not simple and not always accepted by the patient. Moreover, it is not free from risks for the patient health.

Also, we expect that the machines could be simplified and used by the patients at their homes. This will permit a benefit for the patient health and a better recovery of the lost functions.

To reach this medium-term result, a sequel of supportive actions is needed which can be summarized by i) a low-cost system, ii) an easy-to-use system, iii) a remote system control by hospital or rehabilitation institute and iv) a simple maintenance.

The home rehabilitation system must be tailored on the patient needs and should be used for long time. Because it will be used by a large number of patients, its cost must be much lower than the hospitalization cost in order to be supported by national healthcare organizations. What is necessary is to create a new industry for mass production of the hardware of the ICT-based Neuro-motor Rehabilitation systems together with a maintenance structure able to support the patients.

Because the various system modes of operations depend on software, this should be under control of the hospital supervising the patient and it will be loaded at the beginning of the home service. The system needs also an accurate supervision from the hospital and this is a new service that must be carefully designed. It requires a team of specialists in different medical disciplines which takes care of the patients, collects the data from the remote machines and checks the improvements. On this basis, periodically, they change the therapy and have a remote meeting with the patient. The mass database so far generated would permit researchers to verify the efficacy of the new ICT-based therapies and also could lead towards new insights in the brain recovery mechanisms. These studies would also allow neurologists to have a better knowledge of brain neuroplasticity and to propose more efficient rehabilitation procedures. This means that home-based rehabilitation will define new standard protocols.

From engineering point of view of the home system, it must have a friendly interface; that is, it should be used in a simple way. It is a difficult task to design such a proper interface, and it is foreseeable that the help of psychologist will permit us to better define the steps to be followed. Also, the connection of the sensors must be very simple without any cumbersome manoeuvre by the patient. It should be limited to close a strap. In this direction, the use of EMG sensors appears in line with the simplicity, while the use of today's EEG caps seems to be complex and difficult to be used by most of patients.

Instructions how to use the systems and how to wear the sensors should be explained by a short video.

The remote control of the system will guarantee that it will be used properly and that it permits to improve the performance of the patient.

Finally, maintenance service must be organized to check from remote the correct hardware and software system operations. It requires that the system is periodically updated and controlled on viruses or malwares that could misuse sensitive personal data about the health status of the patient.

All the above aspects will require specific design and what we expect is that this kind of system will be available with the complete associated support organization within a decade from now.

In order to let this system start with its service, it is needed that national health organizations introduce it within their package of services they offer to the customers. This is not so obvious because national health organizations need to move from the standard operating way where the patient gets a service at a centralized service point (hospital, doctor specialist, etc.), while in this new vision by using ICT services, we introduce a remote service where patients are at home and get a service from the doctor which is at the hospital or in some other remote location.

In case all the above requirements are solved, trials will begin shortly. What we expect is that the first step is to move the ICT systems used today in hospital to decentralized health service centres. These centres are closer to the patient home, where trained personnel (with less expertise than the ones at the hospital) will use the ICT rehabilitation systems to several patients on day-to-day bases. Next, the new generation of ICT rehabilitation systems will be appropriately re-engineered and its use at home will begin. This step will require about another three to four years.

5.7 Conclusions

Integrated ICT neurorehabilitation systems most probably represent the future of a widespread recovery of motor functions among neurological patients. Today's standard rehabilitation techniques could be performed in highly specialized hospitals/centres, only, where therapies last for several months. However, a limited number of patients can access the most effective rehabilitative programmes and have the opportunity to regain the compromised ADL independence.

The development of the new ICT-based neurorehabilitation systems could allow a larger number of patients to benefit of the most appropriate rehabilitation at an affordable cost. The implementation of these systems at decentralized healthcare units could constitute a first large-scale use of these new products and the first step showing – in five to ten years, roughly saying – the efficacy of the new rehabilitation perspective.

Moreover, the massive data acquired from this new concept will allow neuroscience researchers to acquire further knowledge about brain neuroplasticity and develop more efficient rehabilitative therapies.

In a long-term perspective, once such a new ICT-based rehabilitative system is implemented, we could foresee three phases: during the first period after the injury, namely *acute*, the patient will be hospitalized for a couple of months. Afterwards, the *post-acute* phase, the patient could benefit of the decentralized local unit's daily assistance, close to his/her home, where the specialists can employ devices similar to the ones available at home later on. During this phase, that can last for few months, caregivers and patients learn how to use the neurorehabilitation devices, through advanced courses (the caregivers) or the daily practice (the patients). When the patient enters the *chronic* phase, (s)he should be provided of his/her own neurorehabilitation device at home and continue a highly specific and intense rehabilitative training by him/herself. However, (s)he will be regularly contacted by the personnel of the local unit in order to perform clinical evaluation tests and, if necessarily, to update the rehabilitative programme. Similarly, but with a lower frequency, the patient will be recruited again by the main hospital in order to verify the overall effectiveness of the therapy performed during the self-management period.

References

[1] Vidal, J. J., "Toward Direct Brain-Computer Communications", *Annu. Rev. Biophys. Bioeng.,* vol. 2, pp. 157–180, 1973.

[2] Farwell, L. A., and E. Donchin, "Talking off the top of your head: toward a mental prosthesis utilizing event-related brain potentials, *EEG Clin. Neurophysiol.,* vol. 70, n. 6, 510–523, 1988.

[3] Birbaumer, N., N. Ghanayim, T. Hinterberger, I. Iversen, B. Kotchoubey, A. Kubler, J. Perelmouter, E. Taub and H. Flor, "A spelling device for the paralysed", *Nature,* 1999.

[4] Cisotto, G., S. Pupolin, M. Cavinato, F. Piccione, "An EEG-based BCI platform to improve arm reaching ability of chronic stroke patients by means of an operant learning training with a contingent force feedback," International Journal of E-Health and Medical Communications, vol. 5, pp. 114–134, Jan–March 2014.

[5] World Health Organization (WHO) Statistics (2013) Annual Report, http://www.who.int/gho/publications/world_health_statistics/2013/en/.

[6] Quinn, T. J., S. Paolucci, K. S. Sunnerhagen, J. Sivenius, M. F. Walker, D. Toni, K. R. Lees, and European Stroke Organisation (ESO) Executive Committee, ESO Writing Committee, "Evidence-based stroke rehabilitation: an expanded guidance document from the european stroke organisation (ESO) guidelines for management of ischemic stroke and transient ischemic attack 2008", *Journal of Rehabilitation Medicine,* vol. 41, pp. 99–111, 2008.

[7] Langhorne, P., F. Coupar, and A. Pollock, "Motor recovery after stroke: a systematic review", *Lancet Neurology,* vol. 8, pp. 741–754, 2009.

[8] Langhorne, P., J. Bernhardt, and G. Kwakkel, "Stroke rehabilitation", *Lancet,* vol. 377, pp. 693–1702, 2011.

[9] Kamao, H., M. Mandai, S. Okamoto, N. Sakai, A. Suga, S. Sugita, J. Kiryu, M. Takahashi, "Characterization of human induced pluripotent stem cell-derived retinal pigment epithelium cell sheets aiming for clinical application," *Stem Cell Reports,* vol. 2, n. 2, pp. 205–218, Jan 23, 2014. doi: 10.1016/j.stemcr.2013.12.007.eCollection 2014.

[10] Uchida, K., M. Nakamura, H. Ozawa, S. Katoh and Y. Toyama, *Neuroprotection and Regeneration of the Spinal Cord,* Springer, 2014.

[11] Dimyan, M. A. and L. G. Cohen, "Neuroplasticity in the context of motor rehabilitation after stroke", *Nature Review Neurology,* vol. 7, pp. 76–85, 2011.

[12] Stroemer, R. P., T. A. Kent, and C. E. Hulsenbosch, "Neocortical neural sprouting, synaptogenesis, and behavioral recovery after neocortical infarctionin rats," *Stroke,* vol. 26, n. 11, pp. 2135–2144, 1995.

[13] Carmichael, S. T., "Cellular and molecular mechanisms of neural repair after stroke: making waves", *Annals of Neurology,* vol. 59, n. 5, pp. 735–742, 2006.

[14] Biernaskie, J. and D. Corbett, "Enriched rehabilitative training improved forelimb motor function and enhanced dendritic growth after focal ischemic injury", *Journal of Neuroscience,* vol. 21, n. 14, pp. 5272–5280, 2001.

[15] Adkins, D. L., J. Bochum, M. S. Remple, and J. A. Kleim, "Motor training induces experience-specific patterns of plasticity across motor cortex and spinal cord", *Journal of Applied Physiology,* vol. 101, n. 6, pp. 1776–1782, 2006.

[16] Rayegani, S. M., S. A. Raeissadat, L. Sedighipour, I. M. Rezazadeh, M. H. Bahrami, D. Eliaspour, and S. Khosrawi, "Effect of neurofeedback and electromyographic-biofeedback therapy on improving hand function in stroke patients," *Top Stroke Rehabil.,* vol. 21, n. 2, pp. 137–151, Mar-Apr 2014.

[17] Bauer, P., C. Krewer, S. Golaszewski, E. Koenig, F. Müller, "Functional electrical stimulation assisted active cycling - Therapeutic effects in patients with hemiparesis from 7 days to 6 months after stroke. A randomized controlled pilot study," *Arch Phys Med Rehabil.* 2014.

[18] Ibitoye, M. O., E. H. Estigoni, N. A. Hamzaid, A. K. Wahab, G. M. Davis, "The effectiveness of FES-evoked EMG potentials to assess muscle force and fatigue in individuals with spinal cord injury," *Sensors* (Basel), vol. 14, n. 7, pp. 12598–12622, 2014.

[19] Ono, T., K. Shindo, K. Kawashima, N. Ota, M. Ito, T. Ota, M. Mukaino, T. Fujiwara, A. Kimura, M. Liu, and J. Ushiba, "Brain-computer interface with somatosensory feedback improves functional recovery from severe hemiplegia due to chronic stroke," *Front Neuroeng.,* vol. 7, n. 19, 2014.

[20] Thrane, G., T. Askim, R. Stock, B. Indredavik, R. Gjone, A. Erichsen, and A. Anke, "Efficacy of Constraint-Induced Movement Therapy in Early Stroke Rehabilitation: A Randomized Controlled Multisite Trial," *Neurorehabil Neural Repair.* 2014.

[21] Wolf, A., R. Scheiderer, N. Napolitan, C. Belden, L. Shaub, and M. Whitford, "Efficacy and task structure of bimanual training post stroke:

a systematic review," *Top Stroke Rehabil.*, vol. 21, n. 3, pp. 181–196, May-Jun 2014.

[22] Rossiter, H. E., M. R. Borrelli, R. J. Borchert, D. Bradbury, and N. S. Ward, "Cortical Mechanisms of Mirror Therapy After Stroke," *Neurorehabil Neural Repair*, 2014.

[23] Pollock, A., S. E. Farmer, M. C. Brady, P. Langhorne, G. E. Mead, J. Mehrholz, and F. van Wijck, "Interventions for improving upper limb function after stroke,"*Cochrane Database Syst Rev.*, vol. 11, Nov 12, 2014.

[24] Coupar, F., A. Pollock, F. van Wijck, J. Morris, P. Langhorne, "Simultaneous bilateral training for improving arm function after stroke," *Cochrane Database Syst Rev.,* vol. 14, n. 4, 2010.

[25] Laffont, I., K. Bakhti, F. Coroian, L. van Dokkum, D. Mottet, N. Schweighofer, J. Froger, "Innovative technologies applied to sensorimotor rehabilitation after stroke," *Ann. Phys. Rehabil. Med.*, vol. 57, n. 8, pp. 543–551, Nov. 2014.

[26] Pfurtscheller, G., and C. Neuper, "Motorimagery and direct brain-computer communication," *Proceedings of the IEEE*, vol. 89, n. 7, pp. 1123–1134, 2001.

[27] Wolpaw J. R., N. Birbaumer, D. J. McFarland, G. Pfurtscheller and T. M. Vaughan, "Brain-Computer interfaces for communication and control", *Clinical Neurophysiology,* vol. 113, pp. 767–791, 2002.

[28] Silvoni, S., A. Ramos-Murguialday, M. Cavinato, C. Volpato, G. Cisotto, A. Turolla, F. Piccione, and N. Birbaumer, "Brain-Computer interface in stroke: A review of progress", *Clinical EEG and Neuroscience*, vol. 42, n. 4, pp. 245–252, 2011.

[29] Ang, K. K., and C. Guan, "Brain-Computer Interface in Stroke Rehabilitation," *Journal of Computer Sci. Eng.,* vol. 7, pp. 139–146, 2013.

[30] Grosse-Wentrup, M., D. Mattia, and K. Oweiss, "Using brain-computer interface to induce neural plasticity and restore function", *Journal of Neural Engineering*, vol. 8, n. 2, 2011.

[31] Hochberg, L. R., D. Bacher, B. Janosiewicz, N. Y. Masse, J. D. Simeral, and J. Vogel, et al., "Reach and grasp by people with tetraplegia using a neurally controlled robotic arm," *Nature*, vol. 485, pp. 372–377, 2012.

[32] Collinger, J. L., B. Wodlinger, J. E. Downey, W. Wang, E. C. Tyler-Kabara, and D. J. Weber, et al., "High-performance neuroprosthetic control by an individual with tetraplegia," *Lancet*, vol. 381, pp. 557–564, 2013.

[33] McFarland, D. J., L. A. Miner, T. M. Vaughan, and J. R. Wolpaw, "Mu and beta rhythmtopographies during motor imagery and actualmovements," *Brain Topography*, vol. 12, n. 3, pp. 177–186, 2000.

[34] Tagliabue, M., and J. McIntyre, "A modular theory of multisensory integration for motor control," *Front Comput Neurosci.*, vol. 8, n. 1, 2014.

[35] Piron, L., A. Turolla, M. Agostini, C. S. Zucconi, L. Ventura, P. Tonin, and M. Dam, "Motorlearning principles for rehabilitation: A pilot randomized controlled study in post stroke patients," *Neurorehabilitation and Neural Repair*, vol. 24, n. 6, pp. 501–508, 2010.

[36] Fetz, E. E., "Operant Conditioning of Cortical Unit Activity", *Science*, vol. 163, pp. 955–958, 1969.

[37] Skinner B. F. "The Behavior of Organisms: An Experimental Analysis", Cambridge, Massachusetts: B. F. Skinner Foundation, 1938.

[38] Wyler, A. R., A. A. Ward Jr, and E. E. Fetz, "Operant conditioning of EEG for control of human epilepsy," *Proc. Electroencephalogr Clin Neurophysiol.*, vol. 39, n. 4, pp. 433, Oct. 1975.

[39] Caria, A., C. Weber, D. Brötz, A. Ramos, L. F. Ticini, A. Gharabaghi, C. Braun, and N. Birbaumer, "Chronic stroke recovery after combined BCI training and physiotherapy: a case report," *Psychophysiology,* vol. 48, n. 4, pp. 578–582, Apr. 2011.

[40] Hutchinson, D. T., "The quest for the bionic arm," *J. Am. Acad. Orthop. Surg*, vol. 22, n. 6, pp. 346–351, June 2014.

[41] Bermúdez I Badia, S., M. S. Cameirão, "A Novel Worldwide Accessible Motor Training Approach for At-Home Rehabilitation after Stroke," *Stroke Res Treat.*, 2012.

[42] Oddy, M., and S. da Silva Ramos, "Cost effective ways of facilitating home based rehabilitation and support," *Neuro Rehabilitation*, vol. 32, n. 4, pp. 781–790, 2013.

[43] Bieńkiewicz, M. M., M. L. Brandi, G. Goldenberg, C. M. Hughes, and J. Hermsdörfer, "The tool in the brain: apraxia in ADL. Behavioral and neurological correlates of apraxia in daily living," *Front Psychol.*, vol. 5, pp. 353, 2014.

[44] Stopczynski, A., C. Stahlhut, J. E. Larsen, M. K. Petersen, and L. K. Hansen, "The smartphone brain scanner: a portable real-time neuroimaging system," *PloS One*, vol. 9, n. 2, 2014.

[45] Wascher, E., H. Heppner, and S. Hoffmann, "Towards the measurement of event-related EEG activity in real-life working environments," *Int J Psychophysiol.*, vol. 91, n. 1, pp. 3–9, Jan. 2014.

[46] De Vos, M., S. Debener, "Mobile EEG: towards brain activity monitoring during natural action and cognition," *Int. J. Psychophysiol.*, vol. 91, n. 1 pp. 1–2, Jan. 2014.

[47] http://www.btsbioengineering.com/

[48] http://http://www.khymeia.com/

[49] Shin, J-H., H. Ryu, and S. H. Jang, "A task-specific interactive game-based virtual reality rehabilitation system for patients with stroke: a usability test and two clinical experiments," *J. Neuroeng Rehabil*, Vol. 11, March 2014.

[50] Piovesan, D., P. Morasso, P. Giannoni, M. Casadio, "Arm Stiffness During assisted Movement After Stroke: The Influence of Visual Feedback and Training," *IEEE Trans. Neural Sys. and Rehab. Eng.*, Vol. 21, n. 3, May 2013.

[51] http://www.interactive-motion.com

[52] http://www.tyromotion.com

Biographies

G. Cisotto received M.Sc. in Communications Engineering in 2010 and Ph.D. in 2014 from the University of Padua. She had worked at IRCCS Hospital Foundation San Camillo of Venice (Italy) from 2010 to 2013, focusing on EEG signals processing for amyothrophic lateral sclerosis and the minimal consciousness state patients. Recently, she had mainly worked on Brain Computer Interfaces for motor rehabilitation of stroke patients. From April 2014 she is research associate at Keio University of Yokohama (Japan) and visiting student at the National Centre of Neurology and Psychiatry of

Tokyo (Japan). Her current interests include processing and analysis of EEG and EMG signals, rehabilitation from focal hand dystonia and Brain Computer Interface.

S. Pupolin, graduated in Electronic Engineering from the University of Padova, Italy, in 1970. Since then he joined the Department of Information Engineering, University of Padua, where currently is Professor of Electrical Communications. He was Chairman of the Faculty of Electronic Engineering (1990–1994), Chairman of the PhD Course in Electronics and Telecommunications Engineering (1991–1997), (2003–2004) and Director of the PhD School in Information Engineering (2004–2007). Chairman of the board of PhD School Directors (2005–2007), Member of the programming and development committee (1997–2002), Member of Scientific Committee (1996–2001), Member of the budget Committee of the Faculty of Engineering (2003–2009) Chairman of the budget committee of the Department of Information Engineering (2014–2017), of the University of Padua. Member of the Board of Governor of CNIT "Italian National Interuniversity Consortium for Telecommunications" (1996–1999), (2004–2007), Director of CNIT (2008–2010). General Chair of the 9-th, 10-th and 18-th Tyrrhenian International Workshop on Digital Communications devoted to "Broadband Wireless Communications", "Multimedia Communications" and "Wireless Communications", respectively, General Chair of the 7th International Symposium on Wireless Personal Multimedia Communications (WPMC'04).

He spent the summer of 1985 at AT&T Bell Laboratories on leave from the University of Padua, doing research on Digital Radio Systems.

He was Principal investigator for national research projects entitled "Variable bit rate mobile radio communication systems for multimedia applications" (1997–1998), "OFDM Systems with Applications to WLAN Networks" (2000–2002), and "MC-CDMA: an air interface for the 4th generation of wireless systems" (2002–2003). Also, he Task leader in the FIRB PRIMO Research Project "Reconfigurable platforms for broadband mobile communications" (2003–2006).

He is actively engaged in research on Digital Communication Systems and Brain Communication Interface for motor neuro-rehabilitation.

6

Ethical Issues in the Use of Information and Communication Technologies in the Health Care of Patients with Neurological Disorders

Matteo De Marco[1] and Annalena Venneri[1,2]

[1]Department of Neuroscience, University of Sheffield, Sheffield, UK
[2]IRCCS, Fondazione Ospedale San Camillo, Venice, Italy
Corresponding author: Annalena Venneri <a.venneri@sheffield.ac.uk>

6.1 Introduction

"Fellow executives, it gives me great pleasure to introduce you to the future of law enforcement: ED209".
Omni Consumer Product's senior president Dick Jones, RoboCop, 1987.

The famous scene of the movie in which the prototype of a mechanized droid is presented to the panel and this, due a "glitch", kills one of the board members is a voluntarily exaggerated and unrealistic example of possible unforeseeable incidents associated with the use of technological means in daily-life environments. It is, however, a good starting point to introduce information and communication technology (ICT) which represents today one of the newest, most innovative and helpful instruments at disposal of a large set of fields in everyday life, and to warn about possible ethical issues arising from its implementation and regular use in different settings. The habitual implementation of ICT in primary schools, for instance, offers an invaluable opportunity for raising the standards of teaching processes and exerts its benefits on the many figures composing this world: teaching staff, pupils and students, as well as, indirectly, on their parents or carers. Albeit playing a particularly major role in specialties which have educational purposes, the applicability of ICT elements is not restricted to schools and universities. There are, in fact, other environments which have become fertile grounds for the application of technological expertise, where ICT can contribute to the

Neuro-Rehabilitation with Brain Interface, 121–142.

fulfilment of the environment's main goals. On this note, the world of health care is a vivid example.

6.2 ICT for Health Care

Although a number of opportunities are offered by ICT in medical practice, it has to be acknowledged that both, the concept of ICT and the idea of medical settings, are two heterogeneous realities. For this reason, questioning whether ICT might enhance and benefit the optimization of medical procedures will probably not lead to a unique answer. For instance, nowadays, there are hospital tasks which strongly resist digitalization, such as managing a patient list in a ward (Iversen et al., 2015). In addition, bird's-eye view studies observed how the proportion of overall computerization of medical data or the general incorporation of ICT in different countries is not particularly high (Anderson & Balas, 2006; Vagelatos et al., 2002; Yanusaga et al., 2008). Conversely, the design of technology-based interventions for the rehabilitatory phases of patient management (after a stroke, for instance) is pursued worldwide and in various shapes (e.g. Agostini et al., 2014; Turolla et al., 2013). These contradictory pieces of evidence represent simple examples to introduce the core topic of this chapter. While the great perspectives associated with the use of ICT for health care may be unequivocal, there is another "dark" side of the coin that deserves careful attention. This is represented by the set of ethical problems that arise from the use of ICT. Within the vast category of medical diagnoses, patients diagnosed with neurological and neurodegenerative conditions deserve particular attention for two main reasons. First, the costs associated with disability related to these conditions sustained by health care and families are substantial (Obermann & Lyon, 2014; Schaller et al., 2014) and are destined to increase, since projections of prevalence rates are not encouraging (Brookmeyer et al., 2007; Dorsey et al., 2007). These numbers indicate that any reasonable research and clinical avenue should be considered for testing or implementation in normal routines. Second, patients with neurological and neurodegenerative disorders (including diagnoses of dementia of non-neurodegenerative nature) are characterized by particular traits which differentiate them from other categories of in/outpatients. These features (e.g. lack of awareness manifested by patients with right-sided stroke or patients with minimal Alzheimer's disease, or the state of complete paralysis combined with retained sharp cognition evident in a proportion of patients with motor neurone disease) indicate that it is necessary to focus on ICT and on how specific ethical aspects associated with ICT are relevant for this clinical population. On this note, for instance, the guidelines of the UK National Institute for Health Research (VV.AA, 2014a) distinguish between the primary care of adults, the "Paediatric Setting", and "Adults Lacking Capacity", as three specific categories of participants in clinical experimentation with their own specific regulations. This same rationale has to be transposed to the characterization of ethical problems in association with the use of ICT.

6.3 Ethical Issues Associated with ICT for Health Care

Within the complete process of actively addressing all ethical issues associated with ICT, whether this is carried out in a structured framework (e.g. Harris et al., 2008) or not, a convenient *modus operandi* is to break down the procedure into two steps: the efforts necessary to explore an ICT procedure and identify possible issues, and the attempts to bypass or solve these problems. The European platform "Ethical Issues of Emerging ICT Applications" acknowledges that the technological interface itself does not represent a threat for anybody. Instead, the real issues arise when, in association with the use of ICT, a significant limitation of the people's "capabilities, freedom and choices" is visible (VV.AA., 2014b). For most of routine medical procedures or when they are hospitalized, patients with neurological and neurodegenerative conditions are generally not treated differently from any other "standard" patient. This means that operations such as electronic data entry or practitioner–patient telephonic communications are associated with the same risks, regardless of who the patient is. These risks are all associated with possible breaches of data authenticity, privacy and confidentiality (e.g. Genes et al., 2013; McClintock & Friendship, 1996) and seem to be acknowledged by medical panelists when asked about their willingness to align to these routines (Anderson & Balas, 2006; Haluza & Jungwirth, 2014). Along these lines, a recent survey study focused on the possibility of promoting health via ICT channels, and findings suggest that even such an apparently safe initiative might hide risks for protection of personal data (Haluza & Jungwirth, 2015). This aspect is particularly relevant for the aging population who are at greater risk of developing cognitive impairment due to stroke and/or cognitive decline and dementia. In fact, over the last years, there has been an increasing research interest in the field of lifestyle variables and disease prevention (Di Marco et al., 2014). A scenario of health insurance rates depending "on individual information and communication technology-tracked lifestyle choices", as provocatively fictionalized by Haluza and Jungwirth (2015) in their study, would have a dramatic impact on the patient–caregiver dyad who will have to be referred to a neurological examination in the future. Moreover, a progressive impairment of financial abilities can be observed as a consequence of a stroke or in the earliest stages of major neurodegenerative diseases (Martin et al., 2013; Triebel et al., 2009). Based on this, the well-being of patients and caregivers would be endangered if a third person or a company were taking personal advantage of an accidental or malevolent leak of personal details. In summary, safety of personal data is paramount in the medical setting and should possibly be even more inviolable for patients with neurological and neurodegenerative conditions. As a consequence, major elements of doubt should be always addressed when implementing ICT for the management of this type of information in neurological and geriatric settings. As a conclusive remark, it is worth noting that any ICT failure of this nature goes beyond the mere technicality of the problem itself. This means that the efficiency and reliability of ICT is not just a matter of identifying a potential problem and devising a solution, but it is a necessary standard to safeguard patients' trust (McNeal, 2014). On this note, a breach of patients' anonymity and/or confidentiality would be a very serious issue of public significance, especially if occurred to a population of individuals who deserves a particular level of protection, such as those suffering from neurological and neurodegenerative conditions.

Protection of restricted information is not the only aspect ethics of ICT should focus on. Even in a virtual scenario with incorruptible data safety, there would be additional aspects to be addressed. These do not refer to the ICT "scaffold" of patient management (the methods by which information is managed, such as electronic databases, or telephone-/email-based channels of communication), but rather to what represents the ICT "content" of medical procedures. In the clinical management or in the experimental investigation of neurological and neurodegenerative diseases, two large entities can be identified in association to which patients are primarily involved: procedures concerning *assessment* and procedures concerning *intervention*. ICT may be successfully implemented at both stages, and the nature of the relevant aspects of ethical concern will be completely different from that covered in the previous section. Although older adults are often idealized as individuals who avoid and resist technological progress and are "not compatible" with an active role of ICT usage, this is challenged by some and branded as myth (Wandke et al., 2012). Recent findings suggest, however, that this belief may be at least partially true (Blažun et al., 2014).

6.3.1 State of the Art on ICT for Health Care

The literature on screening/evaluative instruments of neuropsychological signature based on ICT consists of a large number of instruments validated for clinical use (e.g. Heaton et al., 2014) and has been considered as a powerful and useful avenue even from the American Psychological Association (1986). Normally, the use of these tools results from an initial feasibility study carried out on a small sample extracted from a population comparable to that in which the instrument will be applied. This step is generally followed by a validation carried out in a reference population, in order to finally obtain a set of normative data. Although this methodological practice protects the validity and reliability of the resulting instrument, it does not take into account other important aspects. First, laboratory testing does not necessarily meet ecological validity with all populations. This idea can be extrapolated from the framework by Goel (2010), who conceptualized the dissociation between performance on laboratory cognitive testing and performance in everyday life shown by some neurological patients with prefrontal lobe damage, who perform within the normal range on laboratory-based problem-solving tests, but completely fall apart in their daily-life problem-solving. Second, not all elderly adults are equally familiar with and well disposed towards computerized devices (e.g. Blažun et al., 2014). Computer expertise itself does not seem to be, however, a factor influencing systematically the final test scores. In fact, and third factor, attitudinal and motivational factors seem to play a more significant role (Fazeli et al., 2013). Virtually, these aspects could be particularly exacerbated in adults with suspected mild cognitive impairment or dementia. Reasonably, any adult referred for a neuropsychological examination for suspected cognitive impairment may feel tense or anxious prior to and during the assessment. Although state anxiety does not systematically down-regulate levels of cognitive performance, the relationship between the two variables is complicated (Potvin et al., 2013), and there is evidence showing that even the pattern of brain function is affected by in-scanner levels of anxiety, arguably via a modulation of attentional levels (Bishop et al., 2004). Moreover, mood and emotional states can also influence cognitive performance (e.g. Pessoa, 2009; Schuch et al., 2014). Based on this, the implementation of a computerized tool to assess cognitive function might exacerbate the level of discomfort of an anxious adult who may also have depressed mood and who may not have any familiarity with hospitals, laboratories or technology. These sets of circumstances might (even minimally) impact on their performance. A very recent study, however, has shown that this does not necessarily appear to be the case, and ICT-based testing does not suffer from these disadvantages (Canini et al., 2014). In any case, the fact that ecological validity of computerized testing is still being studied today, after several decades of computerization of assessment tools (for a review, see Schlegel & Gilliland, 2007), is a clear indicator of the fact that both the "naturality" of a test as a real-life computational demand and potential ethical concerns for ICT-based cognitive testing are a topic which has not been thoroughly explored as much as their methodological robustness. This is by no means a criticism towards the use of ICT in the neuropsychological assessment of elderly adults, but it aims to be a thought-provoking analysis of the fact that perhaps a more profound consideration is requested when designing and using ICT for the assessment of populations who deserve a particular degree of protection and should be put at ease in such a delicate situation.

The technologizing of screening instruments for people with brain damage has also been put into practice by designing tools and devices which can monitor the evolution of variables relevant for patients' everyday life, such as levels of physical activeness, sleep abnormalities and psychiatric outbursts (Robert et al., 2013). Monitoring is different from assessing, and it should not, therefore, generate a similar distress, supposedly. In addition, many variables that are clinically relevant for neurological and neurodegenerative conditions are only obtainable with the use of ICT, which may, thus, represent a mandatory choice. A recent review was carried out to evaluate how older adults perceive this type of instrumentation in preventing falls. Individual-related intrinsic factors (such as being in control, independence and personal need) and extrinsic factors (usability, among the others) were found to play a crucial role in supporting a positive attitude towards these devices (Hawley-Haque et al., 2014). A robotic device designed to explore a home environment in search of possible hazard which could cause falls also received enthusiastic feedback (Sadasivam et al., 2014). These remarks reflect the personal opinion of older adults and may be considered general principles which should be transposed to any step of the clinical management of this population, including the assessment of cognitive functions.

The second area of ICT "content" of medical procedures for patients with neurological and neurodegenerative conditions is represented by all technological and communicational devices supporting an intervention. A number of ICT sub-categories have been successfully implemented in this field, both as part of clinical routines and as part of experimental studies. The European Union has promoted two ICT "Framework Programmes" (FP) to aid specifically adults with mild dementia and their caregivers cope with the difficulties they face in daily life. The first of the two (FP6), labelled "COGKNOW", is a project which aims to tailor the ICT interface specifically on the dyad's desires and needs in their cognitive functioning, in order to be as user-friendly as possible (Meiland et al., 2007). The second (FP7), labelled "CONFIDENCE", is a framework designed to support activities of daily living in the aging population living in their own accommodation and to minimize circumstances which might cause hazard to individuals. This last project is particularly paradigmatic for this review, because a profound analysis of any possible ethical concern was carried out face to face with every single participant (González-Vega et al., 2011). The "subject-experimenter" collaborative spirit of these frameworks can be considered a gold standard for "content ICT" ethics of projects of clinical interest. The direct opinion of the recruited individual (and their caregivers) might even supersede the point of view of a patented ethical panel (this does not obviously apply to "scaffold ICT" and data protection).

Generally speaking, "content" issues associated with the use of ICT in interventions designed in the field of healthy or pathological aging are limited in comparison with the phase of assessment. While an ICT-based assessment is a critical, "single-shot" interaction with technology, an ICT intervention consists in a prolonged interaction, where a degree of habituation and increased acceptance and expertise is expected over time. The initial degree of acceptance of technology (at recruitment, for instance) depends on a multifaceted set of reasons including personal, social and economical aspects, and expected benefits (Peek et al., 2014). Perceiving ICT as useful during the time of intervention may raise the levels of its acceptance (Callari et al., 2012; Chou et al., 2013).

6.4 Ethical Aspects in Relation to ICT Robotics

One of the most promising ICT fields is robotics. The study of robotic systems in support of various categories of individuals is an ambitious and multidisciplinary enterprise (Schaal, 2008). Robotics applied to the management of neurological and neurodegenerative diseases has led to the design of socially assistive robots. These are machines whose task is to assist a user in activities such as getting up from bed, caring about personal hygiene, and engaging in rehabilitation exercises (Feil-Seifer & Matarić, 2011). These aids have been actively used in dementia care (Huschilt & Clune, 2012; Shibata et al., 2011). The development of a human–robot interaction has led to a profound examination of the ethical aspects surrounding the insertion of a robot in a key position of the life of a human being. Accidents occurred in the robot-aided surgical setting question the responsibilities associated with errors committed by robots (Datteri, 2013). Although not similarly extreme, robot-aided daily activities entail, in a similar way, the identification of responsibilities of the events occurred after the intervention of robotic ICT units. For this and other reasons, specific legislation related to the use of this form of support is needed, since at present the use of health care robots is not legally regulated (Sharkey, 2008). Apart from these possible concerns regarding safety, Sharkey and Sharkey (2006) highlighted the fact that some initiatives in the field of robotics seem to rely too strongly on what they refer to as "natural magic", that is the tendency humans have to believe that a zoomorphic device capable of expressing an emotional state has somehow a degree of consciousness. Based on this, at least part of the mechanisms by which robots should be helpful in health care would be merely based on an illusion (Sharkey & Sharkey, 2006). Such ephemeral mechanism will inevitably abate, and this might contribute to a sense of isolation perceived by the treated adult, as suggested by the European Union (Salvi et al., 2012). The European recommendations read as follows: "Technical solutions should not violate an older person's dignity and it is critical that ICT serves to augment, rather than replace, human interaction". A review focused on the use of assistive robots in the elderly population identified all these aspects of ethical relevance plus an additional one: the stigma that an adult might perceive when being assisted by a robot in a social environment (Zwijsen et al., 2014). All these aspects are extremely important and suggest how ethical control is needed at various levels in the research on robotics.

6.5 Ethical Aspects of Telemedicine

A second major ICT field of interventional research is described by the terms "Telehealth/E-health", "Telemedicine/E-medicine" and "Telecare/E-care" (for the purpose of this chapter, no specific distinction will be made, and the label "Telemedicine" will be used). These disciplines consist of the use of ICT to guarantee "remote" health care for consultation and interventional purposes, most frequently connecting the patient living at home directly with the service. Since an Internet connection and a computer of satisfactory performance is often necessary to provide this service, the telemedicine of the previous decade suffered from equipment cost, transmission fault and insufficient bandwidth issues (Takahashi, 2001). These do not seem to be major problems any longer, as more and more adults have gained competence in computer use and the price of the necessary technology has decreased (Ganapathy, 2005). Fourteen years ago, Cornford and Klecun-Dabrowska (2001) stated that the providers and experimental designers of telemedical services were exclusively concerned with the medical aspects of their cyber-intervention, while, on the contrary, telemedicine should need "to address questions that go well beyond the medical or clinical context and develop from a concern with information society". In a publication featured in the same issue of the same journal, Bauer (2001) asserted instead that it is technological and cost-related justifications that drive the implementation of telemedicine. In both cases, the authors expressed their preoccupations concerning the underestimation of the importance given to ethical aspects. To contextualize telemedicine as an interventional option for preventing or treating neurodegenerative and neurological conditions, it is useful to refer to the framework by Hensel and colleagues (2006). By reviewing how intrusiveness/obtrusiveness of telemedicine impacts on the user, these authors identified eight independent dimensions. Two of these dimensions might be ethically problematic in elderly adults: human interaction and usability. Similar to robotics, an ICT-based intervention where there is no human-to-human interaction might be associated with the risk of inducing a sense of alienation from society (as defined in more general terms by Cornford & Klecun-Dabrowska, 2001). We suggest that this is an aspect that might even resist habituation and might also emerge at later stages of the treatment. On the other hand, the possible issues with usability of the system partially reflect the previously discussed consideration about acceptance of technology among elderly adults. Moreover, it is crucial that ICT user-friendliness and procedural simplicity are maximized. In the field of neurodegenerations, for example, it is well established that patients at the minimal, prodromal or even preclinical stage of a pathological condition leading to dementia very often show significant decline of episodic memory and learning. This might also be the case for patients with focal lesions due to stroke that impair aspects of cognition including memory. Such impairment might completely prevent the patient from learning an "e-routine" if the interface, the instructions and the general structure of the package is not ergonomic, intuitive and sufficiently simple.

6.6 ICT for Improving Patients' Cognitive Functions

A final comment over home-based ICT treatments to improve cognitive functions is necessary. Over the past ten years, there has been an increasing interest in computerized programs to entertain and at the same time improve cognitive function. These have been shaped as videogames or, more professionally, as multimedia tools specifically designed to stimulate cognitive function. This field has generated a significant market, resulting in a "multibillion dollar industry" (Gilbert, 2013). These instruments might be extremely effective for some individuals, but the economic side of the matter raises questions about the possible conflict of interests of those academics involved in these projects, arising from the possible commercial exploitation of these experimental software/platforms. Within their framework describing cyber-management of medication, Palen and Aaløkke (2006) stated that the benefit for the patient associated with the use of telemedicine should always come first, before any benefit for the clinician. The same principle could be transposed to the use of cognitive training telemedicine. Since we currently do not know and cannot experiment directly what the mechanisms are, by which brain structure and brain function are positively regulated by cognitive training (Zatorre et al., 2012), it appears questionable from an ethical point of view that this incarnation of telemedicine is associated with financial profit.

6.7 Ethical Aspects on ICT for Rehabilitation

So far, the focus of the ethical problems arisen from the use of ICT has been mainly on adults experiencing cognitive changes due to senescence, or patients suffering from cognitive impairment caused by neurodegenerative processes. To complete the overview of this chapter, however, it is necessary to refer also to the parallel and similarly important category of patients with focal brain damage. The qualitative nature of the deficits (and thus of the procedures) might be different (e.g. the assessment and treatment of attentional neglect, as in Cipresso et al., 2013) in these patients. The ethical aspects of ICT-based assessment and rehabilitation of cognitive function in this category of patients are substantially the same as those previously presented.

A major additional domain of relevance in these patients is represented by motor deficits. The use of ICT for assessing and treating motor impairment in stroke survivors for instance (but also in patients suffering from multiple sclerosis or Parkinson's disease) exploits the progress and breakthroughs of multidisciplinary projects, and has led to encouraging results (e.g. Shin et al., 2013; Yeh et al., 2014). Reasonably, the ethical points raised to comment on the technologizing of cognitive assessment are less intense when ICT is similarly used to test motor function. Our argument is based on the fact that the use of ICT is probably more suitable for accommodating the performance (or the attempted performance) of a motor act. The interaction between human and machine would generally involve the executive part of the motor act, which is concrete and highly mechanical in nature (probably this does not apply to virtual reality, which, however, is also applied to populations who do not have any diagnosis-dependent motor deficit, e.g. McEwen et al., 2014). On the other hand, the human–machine interaction during cognitive assessment would mostly permeate abstract processes (such as decision making, for instance). For this reason, the use of "hi-tech" equipment would not distort spontaneity of a motor action as perceived by the patient and would thus not trigger any major psychological discomfort (although the possibility that it might intensify the distress perceived because of the inability to perform the requested action cannot be ruled out).

The therapeutic use of ICT for motor impairment deserves instead a deeper analysis. The nature of a motor operation is different from a cognitive process, for a series of reasons which are relevant for the ethical analysis of ICT-based procedures. Failing a cognitive operation that is part of a rehabilitation program will simply result in an unperformed memory or attentional process (with a subjective sense of inadequacy and disappointment, perhaps), but it will never be *life-threatening*. On the other hand, failing a motor operation, especially if part of transfer routines from wheelchair to bed, or walking exercises, might result in a fall and might have extremely dangerous consequences. For this reason, human supervision may be essential when it comes to exercising some motor functions. Moreover, other procedures normally carried out by a physiotherapist, such as manipulation of muscular tissue, are normally performed with extreme care and accuracy, focusing for instance on the exact target area, gauging the appropriate strength (based on the age of the patient, or on diagnostic information, for instance) and adjusting their delivery according to multidimensional feedback (from tactile appreciation of a muscular fibre to, as an extreme example, the sound of a rib cracking). The need for a safe environment, appropriate procedures and expert guidance, unfortunately, implies that hospitalization is often necessary to undergo motor rehabilitation treatment, at least in the earliest stage after stroke. On this note, ICT might give a major contribution in the form of robotics under specialized human supervision. In his detailed published perspective, Schaal (2007) hypothesized a future in which robots might serve as supervisors of motor activities for elderly adults at their home, in an environment they will feel more comfortable with. In this eventuality, there is an important concern which should be carefully addressed. Particular care should be reserved for a profound study of human–robot interaction, as robots should achieve sufficient knowledge "about the neural and musculoskeletal deficiencies of a patient, and tailor exercises and suggestions for improvements accordingly" (Schaal, 2007). From this point of view, robots designed to stimulate communication and cognitive engagement (e.g. Tanaka et al., 2012) are remarkably simpler. Despite the apparent necessity of human expertise in modern-times robotic rehabilitation, there are studies in which ICT devices were implemented for undersupervised or unsupervised home activities, such as joystick-based or wheel-based suites for the upper limb (Johnson et al., 2007) or the wrist (Yamamoto et al., 2014). The transposition of robotic equipment from the laboratory to private premises is a positive achievement for the patient, because they can operate in a familiar and protective environment, but the transfer can only be completed if a program of rehabilitation is judged as completely safe and is thoroughly tailored on an individual's characteristics. Likewise, a relatively simple portable end effector will be likely to be adaptable to a home setting, whereas a complex exoskeleton machine (e.g. Banala et al., 2009) will be not. It has to be acknowledged, however, that motor rehabilitation is going today towards areas of research interest which are based on complex multidisciplinary backgrounds, and it is often very difficult to desupervise a form of motor intervention which is based on devices that request theoretical and practical expertise, such as brain–machine interfaces (Soekadar et al., 2014). Residentialization of treatment may considerably help in easing patients' strain, and in such cases, attention should be paid in order to give further contribution to achieve good tolerance: for instance playing video games as a way to interact with the device while performing the motor exercise components of the training of the upper limb (e.g. Theriualt et al., 2014), or enriching the treatment experience using specific auditory feedback (Rosati et al., 2013).

6.8 Conclusions

Nowadays, the use of ICT is becoming more and more pervasive in medicine, from the use of tablets and smartphones for radiologists (Székely et al., 2013) to mobile phone applications for cognitive screening (Zorluoglu et al., 2014) to Internet-based videoconferencing as a rapid communicational channel between patient and doctor (Mapundu et al., 2012). Ethical aspects of ICT applied to healthy elderly adults and adults with neurological and neurodegenerative problems are a multidimensional area of study which deserves attention and action at multiple levels. For this reason, the literature on ICT and ethical issues is not a unitary body of publications. With specific reference to eHealth, for instance, the published work is highly heterogeneous, poorly indexed, and of inconsistent quality (Car et al., 2008). In this chapter, we presented an overview covering the most important ethical concerns associated with the use of ICT, with particular emphasis on the differentiation between "scaffold" aspects of data safety and protection and "content" aspects related to the use of ICT for assessment and intervention. It is important to note that cross-cultural and cross-national differences in the conception of ICT ethics might exist. For instance, in the USA (where health care is private), identity theft is a crime which is associated with financial damage (VV.AA., 2012). Based on this, a more intricate and detailed legislation should reasonably exist in support of the implementation of ICT. This is not as expected, as in this country data protection is also not as strictly regulated as in Europe (Anderson, 2007). Distinct levels of optimism and confidence shown by practitioners towards ICT may also exist in different countries (Saigí-Rubió et al., 2014). ICT is also associated with cross-cultural differences in its underlying "philosophy". For instance, in Japan, there is a unique cultural conception of robots associated with a large degree of social acceptance (Šabanović, 2014), which should facilitate the study of robotics. Perhaps even for this reason, there is an increasing need for creating international collaborations to answer ethical requests in ICT (Kärki et al., 2013). A recent publication suggested that a convenient formula for a fair ethical assessment of an ICT should consist of workshops involving stakeholders, developers and users altogether, and should precede commercialization (Palm et al., 2013). In fact, as Hofmann (2002) remarked, ICT does not reflect the necessity of decreasing the deal of responsibilities. Actually, responsibilities increase.

References

[1] Agostini M, Garzon M, Benavides-Varela S, De Pellegrin S, Bencini G, Rossi G, Rosadoni S, Mancuso M, Turolla A, Meneghello F, Tonin P. (2014) Telerehabilitation in poststroke anomia. BioMed Research International. Epub Ahead of Print, 15 April 2014.

[2] Anderson JG, Balas EA. (2006) Computerization of primary care in the United States, International Journal of Healthcare Information Systems and Informatics 1(3): 1–23.

[3] Anderson JG. (2007) Social, ethical and legal barriers to E-health. International Journal of Medical Informatics 76(5–6): 480–483.

[4] Banala SK, Kim SH, Agrawal SK, Scholz JP. (2009) Robot assisted gait training with active leg exoskeleton (ALEX). IEEE Transactions on Neural Systems and Rehabilitation Engineering 17(1): 2–8.

[5] Bauer KA (2001) Home-based telemedicine: a survey of ethical issues. Quarterly of Healthcare Ethics 10(2): 137–146.

[6] Bishop SJ, Duncan J, Lawrence AD. (2004) State anxiety modulation of the amygdala response to unattended threat-related stimuli. The Journal of Neuroscience 24(46): 10364–10368.

[7] Blažun H, Vošner J, Kokol P, Saranto K, Rissanen S. (2014) Elderly people's interaction with advanced technology. Nursing Informatics Studies in Health Technologies and Informatics 201: 1–10.

[8] Brookmeyer R, Johnson E, Ziegler-Graham K, Arrighi HM. (2007). Forecasting the global burden of Alzheimer's disease. Alzheimers & Dementia 3(3): 186–191.

[9] Callari TC, Ciairano S, Re A. (2012) Elderly-technology interaction: accessibility and acceptability of technological devices promoting motor and cognitive training. Work 41(Suppl 1): 362–369.

[10] Canini M, Battista P, Della Rosa PA, Catricalà E, Salvatore C, Gilardi MC, Castiglioni I. (2014) Computerized neuropsychological assessment in aging: testing efficacy and clinical ecology of different interfaces. Computational and Mathematical Methods in Medicine. Epub Ahead of Print, 24th July 2014.

[11] Car J, Black A, Anandan C, Cresswell K, Pagliari C, McKinstry B, Procter R, Majeed A, Sheikh A. (2008) The Impact of eHealth on the quality & safety of healthcare. Report for the NHS Connecting for Health Evaluation Programme, March 2008. Accessible from https://www1.imperial.ac.uk/resources/32956FFC-BD76-47 B7-94D2-FFAC56979B74.

[12] Chou CC, Chang CP, Lee TT, Chou HF, Mills ME. (2013) Technology acceptance and quality of life of the elderly in a telecare program. Computers, Informatics, Nursing 31(7): 335–342.

[13] Cipresso P, Serino S, Pedroli E, Gaggioli A, Riva G. (2013) A virtual reality platform for assessment and rehabilitation of neglect using a kinect. Studies in Health Technology and Informatics 196: 66–68.

[14] Cornford T, Klecun-Dabrowska E. (2001) Ethical perspectives in evaluation of telehealth. Cambridge Quarterly of Healthcare Ethics 10(2): 161–169.

[15] Datteri E. (2013) Predicting the long-term effects of human-robot interaction: A reflection on responsibility in medical robotics. Science and Engineering Ethics 19(1): 139–160.

[16] Di Marco LY, Marzo A, Muñoz-Ruiz M, Ikram MA, Kivipelto M, Ruefenacht D, Venneri A, Soininen H, Wanke I, Ventikos YA, Frangi AF. Modifiable lifestyle factors in dementia: a systematic

review of longitudinal observational cohort studies. Journal of Alzheimer's Disease 42(1): 119–135.

[17] Dorsey ER, Constantinescu R, Thompson JP, Biglan KM, Holloway RG, Kieburtz K, Marshall FJ, Ravina BM, Schifitto G, Siderowf A, Tanner CM. (2007) Projected number of people with Parkinson disease in the most populous nations, 2005 through 2030. Neurology 68(5): 384–386.

[18] Fazeli PL, Ross LA, Vance DE, Ball K. (2013) The relationship between computer experience and computerized cognitive test performance among older adults. The Journals of Gerontology. Series B, Psychological Sciences and Social Sciences 68(3): 337–346.

[19] Feil-Seifer D, Matarić MJ. (2011) Ethical principles for socially assistive robotics. IEEE Robotics and Automation Magazine 18(1): 24–31

[20] Ganapathy K. (2005) Telemedicine and neurosciences. Journal of Clinical Neuroscience 12(8): 851–862.

[21] Genes N, Appel J. (2013) Ethics of data sequestration in electronic health records. Cambridge Quarterly of Healthcare Ethics 22(4): 365–372.

[22] Gilbert C. (2013) Effortless brain training: Western hosts TEDx event. London Community News, 8th April 2013. Accessible from: http://www.londoncommunitynews.com/news-story/2520950-effortless-brain-training-western-hosts-tedx-event.

[23] Goel V. (2010) Neural basis of thinking: Laboratory problems versus real-world problems. Wiley Interdisciplinary Reviews: Cognitive Science 1(4): 613–621.

[24] González Vega N, Kämäräinen A, Kalla O. (2011) Ethically inspired care information technology can enable freedom of choice of older users. Sociology Study 1(6): 452–459.

[25] Haluza D, Jungwirth D. (2014) ICT and the future of health care: Aspects of doctor-patient communication. International Journal of Technology Assessment in Health Care 30(3): 298–305.

[26] Haluza D, Jungwirth D. (2015) ICT and the future of health care: Aspects of health promotion. International Journal of Medical Informatics 84: 48–57.

[27] Harris I, Duquenoy P, Jennings R, Pullinger D, Rogerson S. (2008) Ethical assessment of new technologies: A meta-methodology. Journal of Information, Communication and Ethics in Society 9(1): 49–64.

[28] Hawley-Hague H, Boulton E, Hall A, Pfeiffer K, Todd C. (2014) Older adults' perceptions of technologies aimed at falls prevention, detection or monitoring: a systematic review. International Journal of Medical Informatics 83(6): 416–426.

[29] Heaton RK, Akshoomoff N, Tulsky D, Mungas D, Weintraub S, Dikmen S, Beaumont J, Casaletto KB, Conway K, Slotkin J, Gershon R. (2014) Reliability and validity of composite scores from the NIH Toolbox Cognition Battery in adults. Journal of the International Neuropsychological Society 20(6): 588–598.

[30] Hensel BK, Demiris G, Courtney KL. (2006) Defining obtrusiveness in home telehealth technologies: a conceptual framework. Journal of the American Medical Informatics Association 13(4): 428–431.

[31] Hofmann B. (2002) Is there a technological imperative in health care? International Journal of Technology Assessment in Health Care 18(3): 675–689.

[32] Huschilt J, Clune L. (2012) The use of socially assistive robots for dementia care. Journal of Gerontological Nursing 38(10): 15–19.

[33] Iversen TB, Landmark AD, Tjora A. (2015) The peace of paper: Patient lists as work tools. International Journal of Medical Informatics 84(1): 69–75.

[34] Johnson MJ, Feng X, Johnson LM, Winters JM. (2007) Potential of a suite of robot/computer-assisted motivating systems for personalized, home-based, stroke rehabilitation. Journal of Neuroengineering and Rehabilitation 4: 6.

[35] Kärki A, Sävel J, Sallinen M, Kuusinen J. (2013) Ethicted (evaluation process model to improve personalised ICT services for independent living and active ageing)–future scenario. Studies in Health Technologies and Informatics 189: 50–55.

[36] Mapundu Z, Simonnet T, van der Walt JS. (2012) A videoconferencing tool acting as a home-based healthcare monitoring robot for elderly patients. Studies in Health Technology and Informatics 182: 180–188.

[37] Martin RC, Triebel KL, Kennedy RE, Nicholas AP, Watts RL, Stover NP, Brandon M, Marson DC. (2013) Impaired financial abilities in Parkinson's disease patients with mild cognitive impairment and dementia. Parkinsonism & Related Disorders 19(11): 986–990.

[38] McClintock T, Friendship C. (1996) Ethics committees. Information that can be given to researchers over the telephone needs to be clarified. The British Medical Journal 312(7022): 54.

[39] McEwen D, Taillon-Hobson A, Bilodeau M, Sveistrup H, Finestone H. (2014) Two-week virtual reality training for dementia: Single case feasibility study. Journal of Rehabilitation Research and Development 51(7): 1069–1076.

[40] McNeal M. (2014) Hacking health care. Marketing Health Services 34(3): 16–21.

[41] Meiland FJ, Reinersmann A, Bergvall-Kareborn B, Craig D, Moelaert F, Mulvenna MD, Nugent C, Scully T, Bengtsson JE, Dröes RM. (2007) COGKNOW development and evaluation of an ICT-device for people with mild dementia. Studies in Health Technologies and Informatics 127: 166–177.

[42] Obermann M, Lyon M. (2014) Financial cost of amyotrophic lateral sclerosis: A case study. Amyotrophic Lateral Sclerosis and Frontotemporal Degeneration. Epub ahead of print, 23 September 2014.

[43] Palm E, Nordgren A, Verweij M, Collste G. (2013) Ethically sound technology? Guidelines for interactive ethical assessment of personal health monitoring. Studies in Health Technology and Informatics 187: 105–114.

[44] Panel L, Aaløkke S (2006) Of pill boxes and piano benches: "Home-made" methods for managing medication. Proceedings of the 2006 20th anniversary conference on Computer supported cooperative work: 79–88.

[45] Peek ST, Wouters EJ, van Hoof J, Luijkx KG, Boeije HR, Vrijhoef HJ. (2014) Factors influencing acceptance of technology for aging in place: a systematic review. International Journal of Medical Informatics 83(4): 235–248.

[46] Pessoa L. (2009) How do emotion and motivation direct executive control? Trends in Cognitive Sciences 13(4): 160–166.

[47] Potvin O, Bergua V, Meillon C, Le Goff M, Bouisson J, Dartigues JF, Amieva H. (2013) State anxiety and cognitive functioning in older adults. American Journal of Geriatric Psychiatry 21(9): 915–924.

[48] Robert PH, Konig A, Andrieu S, Bremond F, Chemin I, Chung PC, Dartigues JF, Dubois B, Feutren G, Guillemaud R, Kenisberg PA, Nave S, Vellas B, Verhey F, Yesavage J, Mallea P. (2013) Recommendations for ICT use in Alzheimer's disease assessment: Monaco CTAD Expert Meeting. The Journal of Nutrition, Health & Aging 17(8): 653–660.

[49] Rosati G, Rodà A, Avanzini F, Masiero S. (2013) On the role of auditory feedback in robot-assisted movement training after stroke: Review of the literature. Computational Intelligence and Neuroscience, Article ID 586138.

[50] Šabanović S. (2014) Inventing Japan's 'robotics culture': the repeated assembly of science, technology, and culture in social robotics. Social Studies of Sciences 44(3): 342–367.

[51] Sadasivam RS, Luger TM, Coley HL, Taylor BB, Padir T, Ritchie CS, Houston TK. (2014) Robot-assisted home hazard assessment for fall prevention: a feasibility study. Journal of Telemedicine and Telecare 20(1): 3–10.

[52] Salvi M, VV.AA. (2012) Ethics of information and communication technologies. Opinion No. 26, European Union, 2012.

[53] Schaal S. (2007) The New Robotics—towards human-centered machines. Human Frontier Science Program Journal 1(2): 115–126.

[54] Schaller S, Mauskopf J, Kriza C, Wahlster P, Kolominsky-Rabas PL. (2014) The main cost drivers in dementia: a systematic review. International Journal of Geriatric Psychiatry 30(2): 111–129.

[55] Schlegel RE, Gilliland K. (2007) Development and quality assurance of computer-based assessment batteries. Archives of Clinical Neuropsychology 22(Suppl 1): S49–S61.

[56] Schuch S, Koch I. (2014) Mood states influence cognitive control: the case of conflict adaptation. Psychological Research. Epub Ahead of Print, 7th August 2014.

[57] Sharkey N, Sharkey A. (2006) Artifical intelligence and natural magic. Artificial Intelligence Review 25(1–2): 9–19.

[58] Sharkey N. (2008) Computer science. The ethical frontiers of robotics. Science 322: 1800–1801.

[59] Shibata T, Wada K. (2011) Robot therapy: A new approach for mental healthcare of the elderly – A mini-review. Gerontology 57(4): 378–386.

[60] Shin SY, Kim JY, Lee S, Lee J, Kim SJ, Kim C. (2013) Intentional Movement Performance Ability (IMPA): A method for robot-aided quantitative assessment of motor function. IEEE Proceedings of the International Conference on Rehabilitation Robotics.

[61] Soekadar SR, Birbaumer N, Slutzky MW, Cohen LG. (2014) Brain-machine interfaces in neurorehabilitation of stroke. Neurobiology of Disease, Epub Ahead of Print. Doi: 10.1016/j.nbd.2014.11.025.

[62] Székely A, Talanow R, Bágyi P. (2013) Smartphones, tablets and mobile applications for radiology. European Journal of Radiology 82(5): 829–836.

[63] Takahashi T. (2001) The present and future of telemedicine in Japan. International Journal of Medical Informatics 61(2–3): 131–137.

[64] Tanaka M, Ishii A, Yamano E, Ogikubo H, Okazaki M, Kamimura K, Konishi Y, Emoto S, Watanabe Y. (2012) Effect of a human-type communication robot on cognitive function in elderly women living alone. Medical Science Monitor 18(9): CR550–557.

[65] Theriualt A, Nagurka M, Johnson MJ (2014) Therapeutic potential of haptic Theradrive: An affordable robot/computer system for motivating stroke rehabilitation. Proceedings of the 2014 5th IEEE RAS & EMBS International Conference on Biomedical Robotics and Biomechatronics (BioRob) August 12–15, 2014. São Paulo, Brazil.

[66] Triebel KL, Martin R, Griffith HR, Marceaux J, Okonkwo OC, Harrell L, Clark D, Brockington J, Bartolucci A, Marson DC. (2009) Declining financial capacity in mild cognitive impairment: A 1-year longitudinal study. Neurology 73(12): 928–934.

[67] Turolla A, Dam M, Ventura L, Tonin P, Agostini M, Zucconi C, Kiper P, Cagnin A, Piron L. (2013) Virtual reality for the rehabilitation of the upper limb motor function after stroke: A prospective controlled trial. Journal of Neuroengineering and Rehabilitation 10: 85.

[68] Vagelatos A, Sofotassios D, Papanikolaou C, Manolopoulos C. (2002) ICT penetration in public Greek hospitals. Studies in Health Technologies and Informatics 90: 702–706.

[69] VV.AA. (1986) American Psychological Association, Committee on Professional Standards, American Psychological Association, Board of Scientific Affairs, and Committee on Psychological Tests and Assessment, Guidelines for Computer-Based Tests and Interpretations, The Association, 1986.

[70] VV.AA (2012) Creating a Trusted Environment: Reducing the Threat of Medical Identity Theft: Healthcare Information and Management Systems Society: Privacy & Security Task Force June 2012. Accessible from https://www.himss.org/files/HIMSSorg/content/files/Creating aTrustedEnvironment_Reducing_the_Threat_of_Medical_Identify_Theft FINAL.pdf.

[71] VV.AA. (2014a) National Institute for Health Research: Clinical Research Network. Introduction to Good Clinical Practice (GCP). Accessible from http://www.crn.nihr.ac.uk/learning-development/good-clinical-practice.

[72] VV.AA. (2014b) ETICA: Ethical issues of emerging ICT applications. The magazine of the European innovation exchange. Accessible from http://www.etica-project.eu/deliverable-files.

[73] Wandke H, Sengpiel M, Sönksen M. (2012) Myths about older people's use of information and communication technology. Gerontology 58(6): 564–570.

[74] Yamamoto I, Inagawa N, Matsui M, Hachisuka K, Wada F, Hachisuka A. (2014) Research and development of compact wrist rehabilitation robot system. Bio-medical Materials and Engineering 24(1): 123–128.

[75] Yanusaga H, Imamura T, Yamaki S, Endo H. (2008) Computerizing medical records in Japan. International Journal of Medical Informatics 77(10): 708–713.

[76] Yeh SC, Lee SH, Chan RC, Chen S, Rizzo A. (2014) A virtual reality system integrated with robot-assisted haptics to simulate pinch-grip task: Motor ingredients for the assessment in chronic stroke. NeuroRehabilitation 35(3): 435–449.

[77] Zatorre RJ, Douglas Fields RD, Johansen-Berg H. (2012) Plasticity in gray and white: Neuroimaging changes in brain structure during learning. Nature Neuroscience 15(4): 528–536.

[78] Zorluoglu G, Kamasak ME, Tavacioglu L, Ozanar PO. (2014) A mobile application for cognitive screening of dementia. Computer Methods and Programs in Biomedicine. 118(2): 252–262.

[79] Zwijsen SA, Niemeijer AR, Hertogh CM. (2011) Ethics of using assistive technology in the care for community-dwelling elderly people: An overview of the literature. Aging & Mental Health 15(4): 419–427.

Biographies

M. De Marco is a postdoctoral fellow in the Department of Neuroscience at the University of Sheffield, UK. He has been involved in devising and validating a computerised program of cognitive rehabilitation for cognitive impairment due to neurodegeneration which can be transferred and applied to a telemedicine environment.

A. Venneri is Professor of Clinical Neuropsychology in the Department of Neuroscience at the University of Sheffield in the UK, a honorary consultant neuropsychologist at Sheffield Teaching Hospital in Sheffield, UK and she is also the Scientific Director of the IRCCS Fondazione Ospedale San Camillo in Venice, Italy, a neurorehabilitation hospital which uses advanced technologies, robotics and telecare for treating patients with motor and cognitive disorders consequent to stroke or neurodegeneration.

Index

Editor's Biographies

L. Ligthart was born in Rotterdam, the Netherlands, on September 15, 1946. He received an Engineer's degree (cum laude) and a Doctor of Technology degree from Delft University of Technology. He is Fellow of IET and IEEE and academician of the Russian Academy of Transport.

He received Honorary Doctorates at MSTUCA in Moscow, Tomsk State University and MTA Romania. Since 1988, he held a chair at Delft University of Technology. He supervised over 50 PhD's.

He founded IRCTR at Delft University. He is founding member of the EuMA, chaired the first EuMW and initiated EuRAD conference.

Currently he is emeritus professor of Delft University, guest professor at Universities in Indonesia and China, Chairman of CONASENSE, Member BoG IEEE-AESS. His areas of specialization include antennas and propagation, radar and remote sensing, satellite, mobile and radio communications. He has published over 650 papers, various book chapters and 4 books.

R. Prasad is currently the Director of the Center for TeleInFrastruktur (CTIF) at Aalborg University, Denmark and Professor, Wireless Information Multimedia Communication Chair. Ramjee Prasad is the Founder Chairman of the Global ICT Standardisation Forum for India (GISFI: www.gisfi.org) established in 2009.

GISFI has the purpose of increasing of the collaboration between European, Indian, Japanese, North-American and other worldwide standardization activities in the area of Information and Communication Technology (ICT) and related application areas. He was the Founder Chairman of the HERMES Partnership – a network of leading independent European research centres established in 1997, of which he is now the Honorary Chair. He is a Fellow of the Institute of Electrical and Electronic Engineers (IEEE), USA, the Institution of Electronics and Telecommunications Engineers (IETE), India, the Institution of Engineering and Technology (IET), UK, Wireless World Research Forum (WWRF) and a member of the Netherlands Electronics and Radio Society (NERG), and the Danish Engineering Society (IDA).

He is also a Knight ("Ridder") of the Order of Dannebrog (2010), a distinguished award by the Queen of Denmark. He has received several international award, the latest being 2014 IEEE AESS Outstanding Organizational Leadership Award for: *"Organizational Leadership in developing and globalizing the CTIF (Center for TeleInFrastruktur) Research Network"*.

He is the founding editor-in-chief of the Springer International Journal on Wireless Personal Communications. He is a member of the editorial board of other renowned international journals including those of River Publishers. Ramjee Prasad is a member of the Steering committees of many renowned annual international conferences, e.g., Wireless Personal Multimedia Communications Symposium (WPMC); Wireless VITAE and Global Wireless Summit (GWS). He has published more than 30 books, 900 plus journals and conferences publications, more than 15 patents, a sizeable amount of graduated PhD students (over 90) and an even larger number of graduated M.Sc. students (over 200). Several of his students are today worldwide telecommunication leaders themselves.

S. Pupolin, graduated in Electronic Engineering from the University of Padova, Italy, in 1970. Since then he joined the Department of Information

Engineering, University of Padua, where currently is Professor of Electrical Communications. He was Chairman of the Faculty of Electronic Engineering (1990–1994), Chairman of the PhD Course in Electronics and Telecommunications Engineering (1991–1997), (2003–2004) and Director of the PhD School in Information Engineering (2004–2007). Chairman of the board of PhD School Directors (2005–2007), Member of the programming and development committee (1997–2002), Member of Scientific Committee (1996–2001), Member of the budget Committee of the Faculty of Engineering (2003–2009) Chairman of the budget committee of the Department of Information Engineering (2014–2017), of the University of Padua. Member of the Board of Governor of CNIT "Italian National Interuniversity Consortium for Telecommunications" (1996–1999), (2004–2007), Director of CNIT (2008–2010). General Chair of the 9-th, 10-th and 18-th Tyrrhenian International Workshop on Digital Communications devoted to "Broadband Wireless Communications", "Multimedia Communications" and "Wireless Communications", respectively, General Chair of the 7th International Symposium on Wireless Personal Multimedia Communications (WPMC'04).

He spent the summer of 1985 at AT&T Bell Laboratories on leave from the University of Padua, doing research on Digital Radio Systems.

He was Principal investigator for national research projects entitled "Variable bit rate mobile radio communication systems for multimedia applications" (1997–1998), "OFDM Systems with Applications to WLAN Networks" (2000–2002), and "MC-CDMA: an air interface for the 4th generation of wireless systems" (2002–2003). Also, he Task leader in the FIRB PRIMO Research Project "Reconfigurable platforms for broadband mobile communications" (2003–2006).

He is actively engaged in research on Digital Communication Systems and Brain Communication Interface for motor neuro-rehabilitation.